Subjective Well-Being and Social Media

Subjective Well-Being and Social Media

S.M. Iacus

G. Porro

CRC Press
Taylor & Francis Group
Boca Raton London New York

CRC Press is an imprint of the
Taylor & Francis Group, an **informa** business

A CHAPMAN & HALL BOOK

First edition published 2021
by CRC Press
6000 Broken Sound Parkway NW, Suite 300, Boca Raton, FL 33487-2742

and by CRC Press
4 Park Square, Milton Park, Abingdon, Oxon, OX14 4RN

First issued in paperback 2023

© 2021 Taylor & Francis Group, LLC

CRC Press is an imprint of Taylor & Francis Group, an Informa business

No claim to original U.S. Government works

ISBN-13: 978-1-138-39392-9 (hbk)
ISBN-13: 978-1-032-04316-6 (pbk)
ISBN-13: 978-0-429-40143-5 (ebk)

DOI: 10.1201/9780429401435

Publisher's Note
The publisher has gone to great lengths to ensure the quality of this reprint but points out that some imperfections in the original copies may be apparent.

Typeset in LM Roman
by KnowledgeWorks Global Ltd.

Visit the Taylor & Francis Web site at
http://www.taylorandfrancis.com

and the CRC Press Web site at
http://www.crcpress.com

"If there were in the world today any large number of people who desired their own happiness more than they desired the unhappiness of others, we could have a paradise in a few years."
Bertrand Russell

Contents

Preface

This book presents an overview of the most recent projects on the estimation of subjective well-being through social media data.

In particular, we focus on a new project, aimed at constructing a Twitter Subjective Well-Being Index, which started in 2012 - almost at the same time of expansion of sentiment analysis to Twitter data - and grew slowly till the present days.

The project was originally conceived at the University of Milan (Italy) and then embraced later in 2015 by the University of Insubria (Como, Italy), the University of Tokyo and the University of Waseda in Japan.

In the first chapter of the book we review the different approaches to the estimation of well-being, from traditional macro-economic definition - both one-dimensional and multidimensional - to survey analysis and finally to big data and social networking sites (SNS) in particular. The reasons that led us, among other scholars, to focus on SNS as data sources for the evaluation of subjective well-being is not simply the increasing availability of a huge and continually updated flow of texts that the SNS provide. Social media data, in fact, also exhibit interesting methodological advantages, with respect to traditional data sources. For example, they give the social scientist and the policy makers the opportunity to know in almost real-time what people think about (or how they perceive) their quality of life, minimizing the interaction between the researcher and the observed individuals, and thus circumventing the well-known *observer bias* induced by surveys and questionnaires. Moreover, a continuous monitoring of the well-being can help to disentangle the structural component from the extemporaneous one as well as any seasonality effect.

The second chapter, although being quite technical, introduces briefly the most commonly used machine learning and statistical techniques for textual analysis. It serves two scopes: to explain how machines transforms text into meaningful statistics, and also to convey the idea that human supervision is an essential step of this process whatever technique is used. To rephrase Gary King: nowadays, *social science needs to be computer-assisted but has to be human-empowered*. This chapter is accompanied with computer code to ease the understanding of the topics.

The third chapter presents different SNS-based subjective well-being indexes that have been proposed in the literature, with a special focus on the one proposed by the authors of this book. A cross-country comparison between Japan and Italy shows that the perceived well-being is a function of structural conditions (such as wealth and income of a country, ageing, etc.) but also of cultural characteristics of each country, meaning that a single global model of well-being is a bit too ambitious as a goal. A causal inference analysis based on structural equation modeling will shed light on the relative importance of macro-economic and health variables on well-being.

Among all positive aspects of SNS data, there are also some pitfalls which are quite easy to imagine, and well known to the experts in the field. The main one is that social media accounts/users/data cannot be considered statistically representative of the demographic population. In the fourth chapter we present a possible approach to tackle the selection bias problem by anchoring social media indexes to official statistics. Through the application of small area estimation methods, it will be shown how to stabilize social media statistics to produce indicator that may eventually be used to complement official statistics.

The last chapter focuses on the analysis of the impact of the COVID-19 pandemic, that hit the world in 2020, on the social media indexes of subjective well-being previously discussed in the book. While some indicators only show negative peaks for a very short time, the new index proposed here seems to indicate that a prolonged stress condition can indeed impact the structural component of the subjective well-being. In particular, it is shown, through time-series analysis, that the level of subjective well-being dropped dramatically both in Japan and Italy with the start of the pandemic and then remains persistently low for quite a long time. At the time this note is written, we cannot evaluate yet whether the initial pre-pandemic well-being level is going to be recovered and how much time the recovery may require. Through a data science approach, it is also shown that the complexity of the well-being can be captured only through a dynamical analysis that show which mix of factors can explain or impact the subjective well-being through time. Finally, a structural equation model is applied again to attempt some causal analysis on the determinants of subjective well-being as measured by our indicators.

This book comes with two sets of R scripts and data: the first one allows to replicate some of the analyses presented in Chapters 3, 4 and 5, and the second one shows the implementation of the techniques presented in Chapter 2. All the data and scripts will be available at: `https://github.com/siacus/swb-book`.

This work is the outcome of a very long journey along which we had the opportunity to collaborate with several colleagues. Among them, we would like to thank those who contributed very closely with us in developing some aspects of this research: Elena Siletti, Silvia Salini, Matteo Curti, Tiziana Carpi and, on the other side of the world, Nakahiro Yoshida and Airo Hino. We also thank Vasiliki Voukelatou for carefully reading and discussing preliminary versions of some chapters.

We also acknowledge the Japan Science and Technology Agency CREST (Core Research for Evolutional Science and Technology) project, grant n. JP-MJCR14D7, which made possible to acquire the Twitter data for Japan. We also need to thanks the Waseda Institute of Social Media Data (WISDOM) for its support on the human coding of the Japanese tweets and the Data Science Lab of the University of Milan for same activity for the Italian data.

Milan, Como & Tokyo, April 27, 2021

1

Subjective and Social Well-Being

1.1 Introduction

"We understand that this paper raises difficulties for much of well-being research as currently conducted and proposes no solution to those difficulties. Yet, it is better to be aware of the problems than to ignore them. And knowing about them is the first step to finding ways of neutralizing them and doing better" (Deaton and Stone, 2016).

With these words Angus Deaton - Nobel Prize for Economics in 2015 - and Arthur A. Stone conclude one of their studies on well-being evaluation. A similar awareness - *si parva licet componere magnis* - leads the authors of the following pages; we will describe the problem of estimating subjective well-being and emphasize some of the obstacles that the studies on subjective well-being have met in their attempt to carry out an evaluation. Then, we will provide a review of new evaluative techniques based on the availability of large scale data coming from the *Social Networking Sites* (SNS). In this context, we will put emphasis on a proposal made by the authors of this book. Not all the facets of well-being will be taken into consideration, and the number of issues raised will likely overcome the number of problem solved; nevertheless, our aim is to show how this *red*new data can improve the estimation of subjective well-being, mitigating the critical aspects of the evaluation process we will point out.

1.1.1 Subjective Well-Being

First of all, we have to define the object of our study. The well-being - whatever the content given to the term - of a community and its members is (or should be) one of the main concerns of every social analyst and policy maker. In fact, what we wonder when a policy is adopted, an innovation is introduced or an unexpected event happens is: are we - as individuals or as a community - now more or less happy than before? It is not surprising, therefore, that the measurement of both *personal* (or *individual*) and *social* (or *collective*) well-being has captured the attention of psychologists, social scientists and policy makers (Frijters et al., 2020).

These two dimensions of well-being are usually defined as *subjective*: on the one hand, personal well-being concerns the subjective feelings of an individual and the evaluation of his/her own quality of life; on the other hand, social well-being is a synthetic and significant description of how the development of a socio-economic system is perceived as satisfactory, well balanced and sustainable by its inhabitants.

In psychological literature, subjective well-being has been presented as a multifaceted concept. According to some scholars (Lyubomirsky, King, and Diener, 2005) the notion of subjective well-being includes two dimensions: *life satisfaction*, usually defined in relation to different life domains, and *mood*: the first is defined as the evaluative or cognitive component of well-being, while the second as the emotional component (Diener, 1984, Diener, Suh, Lucas, et al., 1999). Other studies (Ryan and Deci, 2001, Deaton and Stone, 2013, Steptoe, Deaton, and Stone, 2015) give a definition of subjective well-being as threefold: well-being has an *evaluative* domain, that coincides with life satisfaction, a *hedonic* one, covering feelings of happiness, sadness, anger, illness and pain and finally an *eudaimonic* dimension, referring to sense of purpose and meaning in life.

The main variable that allows for a distinction among these dimensions is the reference time frame used to express an evaluation of well-being: emotional well-being is related to short-run feelings, life satisfaction more likely assumes a medium or long-run horizon (because one is supposed to answer questions such as: am I satisfied about the life I am living or about the life I have led so far?); the eudaimonic well-being is not only a long-run dimension, but has also a forward-looking perspective (because, as the psychologist Carl R. Rogers wrote, "the good life is a process, not a state of being" (Rogers, 1961)). To some extent - borrowing categories from time-series analysis - we may say that these definitions help us disentangle the "structural" and the "volatile" component of subjective well-being.

1.1.2   Objective Measures

The evolution of subjective well-being evaluation is summarized in Figure 1.1. The definition and measurement of well-being had originally to face a not simple relationship with the objective indicators of welfare and socio-economic development. In economic analysis, in fact, social welfare is often related to the level of economic development which, in turn, is traditionally measured by the gross domestic product (GDP). A possible juxtaposition of well-being to GDP has shown its limits quite soon. Alternative measures of well-being seem, in fact, not to have high and systematic correlation with the evolution of domestic production (see Section 1.2). In this sense, GDP cannot be considered the first indicator of well-being - much less of its subjective component - but rather a precursor and, nowadays, a component of the wide range of indicators developed in socio-economic research.

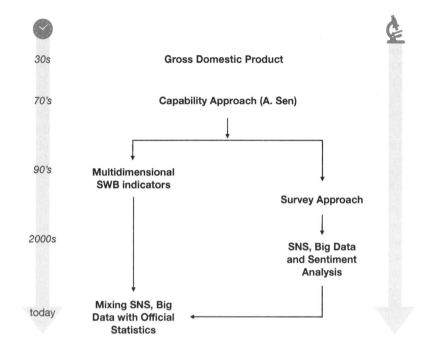

Figure 1.1
How research and literature about well-being evolved through time.

1.1.3 Multidimensional Indicators

Simon Kutznets (1934) and Amartya Sen (1980) - both Nobel laureates in economics - taught that personal and social welfare cannot be simply reduced to the amount of goods and services produced by a given country in a given period of time. This awareness has led to the creation of a bunch of multidimensional well-being indicators, with an increasing focus on the subjective perspective for two reasons (see Section 1.3):

a) among the dimensions that are assumed relevant to well-being, a number is connected to individual perceptions of life quality (including, e.g., freedom of choice);

b) the relative importance of each dimension in determining individual well-being is considered, in turn, subjective; as a consequence, in some cases a dashboard of variables (instead of a single measure) is provided and the weighting and aggregating step is left to the users' choice and preference.

The multidimensional approach to well-being measurement is characterized by some flaws that have been extensively discussed in the literature: first of all, drawing up an exhaustive list of well-being components and evaluating their weight in overall well-being is a challenging task; secondly, several of these dimensions are hardly observable and, at best, can be measured by objective proxies whose soundness is often debatable. Criticism to the approach has come to question its empirical relevance and opened doors to a U-turn change in the strategies for well-being evaluation. If both one-dimensional and multidimensional measures are unreliable, due to the limits of observable variables, the only feasible option to estimate individual and collective well-being is to explicitly *ask people* to express an evaluation about their own condition.

1.1.4   Surveys

To this aim, surveys and questionnaires have been increasingly and widely used to collect information about well-being levels and dynamics of individuals and communities (see Section 1.4). Different methods to conduct surveys have been developed - also conditioned by the technology applied (face-to-face interview, telephone, internet) - in order to disentangle the incidental, emotional aspect of self-reported well-being and the evaluation of life satisfaction, which requires to examine current and past events in a medium or long-run perspective.

But the most significant drawback of survey-based research is the bias induced in well-being evaluation by the survey itself. It is a sort of "Hawthorne effect"[1] that Angus Deaton (Deaton, 2012, Deaton and Stone, 2016) pointed out. In fact, changing the order of the questions of a survey may be sufficient to affect the evaluation the respondents give about their own mood or quality of life. More generally, when the respondents are aware of being asked for an assessment of their own life and of being observed while giving the evaluation, the answer they give may be biased by this awareness. Therefore, the dilemma the analysts face is quite clear: on one hand, they wish *to ask* people for a self-evaluation of their well-being, in order to overcome deficiencies due to measurements based only on observable quantities; on the other hand, they should *not to ask* people for a self-reported evaluation, in order to avoid biases due to the awareness of respondents.

1.1.5   Social Networking Sites and Data at Scale

With the beginning of the era of virtual communication, a new source of large-scale dataset is provided that seems to address the need for such an information; in fact, the availability of a huge and continually updated conversation flow on SNS theoretically provides a real-time opportunity to know what people think about the quality of their own daily life - both from an emotional and an evaluative perspective - without submitting any explicit questionnaire. This is fostering a stream of studies whose aim is to extract

---

[1] Also known as "observer effect", it is the phenomenon by which individuals modify their behavior in response to their awareness of being observed (Landsberger, 1958).

meaningful information from the enormous amount of words or images posted on well-known platforms such as Facebook, Twitter, Instagram (Voukelatou et al., 2020) (see Section 1.5).

The first issue raised by this new data source is a technological one: the increase in computational power of technological devices does not guarantee, *per se*, the ability to separate helpful information from background noise in virtual conversations. Fortunately, recent advancements in statistical theory and its applications are improving the capacity of social scientists to analyze the content of these large-scale dataset and promoting the dissemination of different methods of so-called *sentiment analysis*.

As said, one of the main *advantages* of large-scale datasets coming from SNS is their continuous updating. It offers the opportunity for *nowcasting*[2] activity; in fact, while variables that are more traditionally assumed to be related to welfare - such as GDP or morbidity rates - are observable only with a time lag, that sometimes makes the policy maker intervention less effective, SNS data allow for a real-time monitoring of public sentiment and can anticipate changes in objective variables. Moreover, when the methods for sentiment analysis are language independent (i.e., they can be applied to texts expressed in different languages, without any particular limitation), a comparison among linguistic and socio-cultural contexts is made possible, where not only differences in the use of language can emerge, but also cultural specificities - such as social conventions that impose a more strict self-control in expressing emotions - can be discovered, as far as they are recorded in virtual conversations.

On the other hand, SNS data have also some intrinsic *limitations*. First of all, users of these platforms are not a representative sample of the whole population; therefore, any social well-being evaluation achievable from the analysis of these data cannot be immediately extended to the whole population. Adjusting procedures can be applied to make the results more general but, above all and despite their limited representativeness, SNS can be considered a sort of opinion-making arena, where expressed ideas affect or anticipate collective sentiment and trends. This, actually, suggests that a second drawback can be imputed to evaluation of well-being via SNS data: the use of social networks itself can alter self-perceived or self-declared well-being. In fact, even if SNS users do not answer any explicit question about their own personal status, they are aware they are sharing their feelings with a community and this can distort their well-being self-report, in order to satisfy self-representation needs. Furthermore, SNS messages and texts seem more suitable to reflect short-term mood changes than a long-term evaluation of life quality; therefore, a well-being indicator based on SNS data would be more reliable as a measure of emotional well-being than a source of life evaluation. However, despite the validity of this remark, adequate statistical analysis can help in separating

---

[2]The term is a contraction for "now" and "forecasting": it refers to the opportunity to collect information about some quantities or variables in real time.

the volatile and the structural components of well-being path described by virtual conversations on the net.

1.1.6   What You'll Find (and What You'll Not) in This Book

An extensive body of literature has been produced in all human and social sciences (philosophy, anthropology, psychology, sociology, economics, biology, medicine, public health, etc.) with the aim to define well-being, its content and its relation to happiness. Equally large is the number of studies on the instruments and policies that can be implemented to foster individual and social well-being. This book will make use of some of these definitions and refer to some of these studies, but its scope is not to make a critical survey of the concepts of well-being adopted in different scientific contexts and re-search fields. The topic of the book is the *measurement* of well-being and, in particular, the measures of well-being based on self-declared evaluations of individuals, i.e., what is commonly conceived as subjective well-being.

Our main goal, working on this book, is to provide social scientists, an-alysts, scholars and students an update of the latest SNS-based indexes and projects but a also a new technique and a new instrument to evaluate subjec-tive well-being, focusing on self-perceptions and sentiment daily reported by each of us on social network sites. That is the reason why, in the next pages, we will come back to a brief examination of the content and the historical evolution of the term, in order to provide a better-structured definition of subjective well-being and disentangle possible misinterpretations. The main focus of the book, however, remains the analysis of methodologies and tools used to evaluate subjective well-being in its multiple facets, and the exam of advantages and disadvantages of these methods and instruments.

The new methodology we propose is an algorithm for sentiment analy-sis - named *iSA* (integrated Sentiment Analysis) - that allows for extracting meaning from SNS texts without relying only on automatic classification rules but, on the contrary, assigning a crucial role to human supervision (Ceron, Curini, and Iacus,, 2016). The instrument elaborated for evaluating subjective well-being is the result of the application of iSA to large-scale datasets coming from Twitter: a high-frequency and geo-localized composite indicator of sub-jective well-being, named *SWB* or *Subjective Well-Being* Index. Actually, the acronym may stand for both "subjective" and "social" well-being index; the SWB, in fact, directly evaluates subjective well-being at a collective level, with-out passing through the individual level evaluation, hence circumventing the consequent problem of aggregation of individual preferences (see Section 1.6).

Examples of empirical application of SWB for Italy and Japan (see Chap-ters 3 and 5) will illustrate the features and potential of the indicator in separating structural and volatile component of well-being, forecasting in real time the impact of macro-economic shocks on well-being, disentangle inter-cultural differences. We will also show how the information coming from SWB can helpfully integrate official statistics and provide a more detailed knowledge and monitoring of well-being and its dynamics, due in particular to its frequent refresh and updating (see Chapter 4). At the same time, the examples will make clear how and to what extent the application of iSA to SNS large-scale datasets may overcome some of the obstacles usually met in subjective well-being estimation. Also, limitations of the technique and of the instrument will not be swept under the rug.

In view of the above, the reader will not find in the book a complete survey of the boundless literature either on well-being or even on subjective well-being only. Likewise, neither an exhaustive survey of the methods nor of the instruments created to measure subjective well-being will be attempted. A selection will be made, aimed at illustrating the progress of social science in estimating subjective well-being and the significant step forward that the availability of SNS datasets and, particularly, the application of iSA algorithm to these data allow for.

As said, our target reader is any social scientist, scholar or student who want to learn about evaluating subjective well-being in the era of the so called "big data" or "innovative data" or "alternative data" sources. In order to understand the methodologies and the rationale of the measurement tools described, the knowledge provided by a basic course in statistics or econometrics is highly recommended, but is enough.

1.1.7 Wellbeing, Well Being or Well-Being?

The noun well-being[3] is often spelled as "well being" or "wellbeing". Apparently, the word pairing "well-being" has been in use since the 16th century. According to a simple analysis on the titles of books published in English since 1800 and till the year 2000, it seems that well-being is not only the authentic, old and correct way to mention this topic, but also quite trendy (see, Figure 1.2). In this book, we adopt then the "well-being" version.

1.2 Gross Domestic Product

For decades, the gross domestic product (GDP) - i.e., the market value of final goods and services produced in a given time period - and its growth rate have been the most common indicators of a country's socio-economic development. This also made GDP an index of social welfare, despite its well-known shortcomings as an indicator of well-being: in 1934 Simon Kutznets, Nobel Laureate in 1971 and creator of the modern concept of GDP, said at the US Congress - at that time he was chief architect of the United States

---

[3]An extended discussion can be found here: https://writingexplained.org/wellbeing-or-well-being

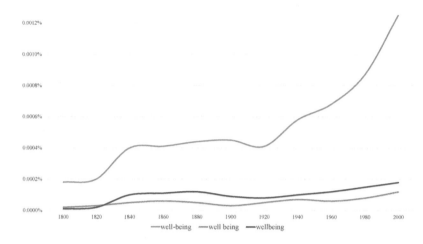

Figure 1.2
Wellbeing, Well being and Well-being used in the titles of books published in English since 1800 till the year 2000. Data source: https://writingexplained.org/

national accounting system - that *"the welfare of a nation can scarcely be inferred from a measurement of national income"* (Kuznets, 1934).

The reasons why GDP had such a success are its capacity to connect goods and services of different nature thanks to monetary evaluation (Stiglitz, Sen, and Fitoussi,, 2009) and the (generally false) assumption that GDP measurement is backed by a linear methodology, objectivity and clearness (for example in public debates) and hence can be particularly useful in international comparisons. In fact, GDP reveals several limitations as a welfare indicator, both on the measurability and on the comparability side. Consider, for instance, that all goods and services that are not traded on the market (e.g., self-handled services for family use or intended for self-consumption) are not included in the GDP estimate.

Specific issues have been raised about the adequacy of GDP as a well-being indicator:

- as said, monetary evaluation of market transaction is the starting point in measuring economic performance and so prices are fundamental, but they may not exist for some goods or services or they may not represent the real value for the society. As a consequence, only goods and services that are officially exchanged on markets are included in GDP calculations, while others, relevant in achieving higher levels of well-being, are overlooked. It is the case of public goods provided by governments like security, freedom and democracy, but also volunteering activities and social relations, health and longevity and all the spectrum of non-market, informal or shadow economies, such as housework and services provided by family members;

- as remarked by Stiglitz, Sen, and Fitoussi, (2009), the change in quality of goods and services is a crucial aspect of the achievable well-being, but it is extremely difficult to measure, because it is generally complex, multidimensional and, once again, to some extent subjective;

- the effect of imperfect competition: firms with a significant market power can rise prices, generating a loss of consumer surplus, compared to the perfect competition equilibrium (Stiglitz, Sen, and Fitoussi,, 2009). The GDP evaluation notices the rising in prices, while the consumers' welfare reduction is disregarded;

- aggregate production measures do not take into account negative externalities such as pollution and the depletion of natural resources and assets like environment and biodiversity;

- using GDP as a proxy of well-being can lead to misleading assessment of welfare levels. On the one hand, negative events such as natural disasters, earthquakes or floods reduce the wealth of societies but may increase their GDP; an estimate of net positive effect of these events on well-being would be highly debatable. On the other hand, aggregate - or even average (e.g., per capita) - measures of national income do not take into account the distribution of income; therefore, a possible disparity in opportunities would not be recorded.

What particularly matters to our aims is that the correlation of GDP with indicators of living standards - despite being high and significant in many cases - is not universal (Stiglitz, Sen, and Fitoussi,, 2009) and differences in income explain only a low proportion of differences in happiness among individuals (Frey and Stutzer, 2002). Amartya Sen (2003) warned that focusing on economic growth and on GDP as measures of human development is misleading, since there is not a one-to-one correspondence between growth and quality of life and too much emphasis on GDP as a unique benchmark of well-being can lead to wrong evaluations and policy decisions. GDP is a measure of market production, more useful to appreciate the dimension of economies than the living standards of citizens (Stiglitz, Sen, and Fitoussi,, 2009). For example, according to Frey and Stutzer (2002), since the Second World War the GDP growth trend is positive in many countries, while the self-reported subjective well-being trend is not.

The overconfidence in GDP comes into conflict with puzzles and paradoxes. The well-known Easterlin paradox (Easterlin, 1974) points out that, at a point in time, happiness varies directly with per capita income both among and within countries; over time, on the contrary, happiness does not grow as income continues to increase. The paradox has been generalized into a "happiness puzzle" that, besides the Easterlin paradox (happiness remains constant as income increases), contemplates the "happiness inertia" (happiness grows, but less than proportionally, when income increases) and the "happiness reversal" (happiness decreases as income increases) cases (Choudhary et al., 2012). It is worth noting that a possible negative relationship between GDP and well-being is also outlined by anti-consumerist theories focused on economically sustainable processes, such as the "degrowth theory", that finds in Serge Latouche one of the main exponents (Latouche, 2006).

Some studies argue that the flattening of happiness trend occurs after some minimum level of income and, even more finely, that the relationship crucially depends on the definition of well-being. Focusing on subjective well-being, for example, Kahneman and Deaton, (2010) find that per capita income positively correlates with life evaluation even at high income levels, while the positive correlation between income and emotional well-being becomes non-significant beyond an annual income of about $75,000. The authors summarize the evidence writing that "high income buys life satisfaction but not happiness".

Several attempts to explain the paradox have been made. Some of them assume that people tend to compare their own behavior to those of their neighbors; hence, at high income levels, their perceived well-being does not grow if their peers show the same (or higher) gain in income (Blanchflower and Oswald, 2004). Other attempts refer to habituation: people get accustomed to their living standard and require a continuous growth of income to maintain their happiness level. Moreover, the Easterlin paradox has been criticized and the absence of positive relationship between income and well-being is not universally accepted.

It follows from the above that GDP is, for sure, an important means to well-being and freedom, but it cannot be considered an exhaustive indicator of well-being, without being potentially misleading for analyses and public policies (Fleurbaey, 2009). In fact, it can only be used as a mere proxy for what really matters for people:

"Yet the gross national product does not allow for the health of our children, the quality of their education or the joy of their play. It does not include the beauty of our poetry or the strength of our marriages, the intelligence of our public debate or the integrity of our public officials. It measures neither our wit nor our courage, neither our wisdom nor our learning, neither our compassion nor our devotion to our country, it measures everything in short, except that which makes life worthwhile".

Robert F. Kennedy at the University of Kansas, March 18, 1968[4]

Although economists and social scientists have tried to ameliorate GDP as a welfare indicator, adding or subtracting monetized aggregates reflecting

---

[4]Robeyns (2005).

different kind of quantities and dimensions (Nordhaus and Tobin, 1973), alternative and complementary approaches to the study and measurement of well-being and new indicators of social welfare have been developed across the years; they have become a core issue in the public debate and a primary concern for policy makers.

That is the reason for the increasing interest of economists, psychologists and philosophers in both multidimensional indicators and self-reported measures of well-being, their significance and their usefulness in policy making (Deaton and Stone, 2013). And that is why we are going to examine them in the following sections.

1.3 Well-Being as a Multidimensional Notion

In 2008, the French government created a commission of inquiry, known as Stiglitz Commission - which, in addition to Joseph Stiglitz, included Jean-Paul Fitoussi and Amartya Sen - with the task of examining how the wealth and social progress of a nation could be measured. In its final report in 2009, the Commission observes that the GDP should not be dismissed in the evaluation of social well-being, but proposes to build a complementary statistical system, suitable for measuring economic sustainability, composed by a wide set of indicators,[5] quantitatively measurable and representing both objective and subjective assessment of well-being, also including - which is particularly interesting to our aims - people's perception of their own quality of life. The Commission identifies key aggregated dimensions that should be taken into account: material living standards, health, education, personal activities including work, political voice and governance, social connections, environment and insecurity. With its work, the Commission made a sort of "paradigmatic" choice that had a strong influence in further well-being literature and, above all, measurement practice.

1.3.1 The Capability Approach

The Stiglitz Commission Report has been preceded - and, to some extent, inspired - by the development of the *capability approach* to the problem of individual and social welfare, originally proposed by Amartya Sen. The approach - which is a broad normative framework for evaluation and assessment of individual well-being, the design of policies and proposals of social changes (Robeyns, 2000, 2006) - covers all the dimensions of human well-being; in

---

[5]Including GDP, "because no single measure can summarize something as complex as the well-being" (Stiglitz, Sen, and Fitoussi, (2009)). Moreover, they argued that synthetic indices could lead to a loss of information, as well as arbitrary assumptions in the weighting that has to be applied to the different components to come to a single index figure.

particular, it devotes much attention to the relationship between material, mental and social well-being. One of the main features of the approach is the distinction between means and ends of human action. Such a distinction introduces two central concepts: *functioning* and *capabilities*. Functioning are more directly related to living conditions, and more empirically measurable, since they concern different aspects of living (Sen and Hawthorn, 1988). Differently, capabilities are possibilities, chances, alternative combinations of functioning that one can choose in order to pursue his/her own aims. Capabilities are thus vectors of functioning, reflecting the person's freedom to lead one type of life or another (Sen, 1992, 2008), i.e., the opportunity to combine different functioning to achieve specific results. For example, reading is a functioning, while reading the content of books contained in a library is a capability, because it implies the opportunity to access the library and the freedom to choose to do it.

A slightly different version of the capability approach has been developed by Martha Nussbaum. The notions of capability are similar in that, also in Nussbaum's works, freedom of choice requires for sure a formal availability of some basic liberties and the assurance of certain levels of material conditions and real circumstances. On the other side, the two concepts are different in that, defining capabilities, Sen refers more frequently to a real or effective opportunity, while Nussbaum pays more attention to individual skills and personality traits (Robeyns, 2005).

In both versions, the capability approach is based on a view of living as a mix of beings and doings, where the quality of life can be defined in terms of valuable functioning and capabilities. This goes duly beyond the "commodity fetishism" and the mere attention to materiality. Nevertheless, the perspective - particularly in the works by Amartya Sen - reminds us of the importance of material conditions and other objectively measurable dimensions of living in evaluating well-being. This will be taken into account when we examine the determinants of subjective well-being, that cannot be reduced to psychological or immaterial circumstances.

1.3.1.1   Empirical Limitations of the Capability Approach

Although the capability approach received great attention by researchers, policy makers and public actors and even if it represents a realistic alternative to traditional methods (such as measurement of income or cost benefit analysis), its applicability is still an issue in the debate (Robeyns, 2006) and doubts concerning the possibility to make empirical use of this "richer but more complex procedure" persist (Sen, 2008).

A relevant empirical question refers to the selection of capabilities and functioning: what kind of capabilities and functioning are valuable in order to assess well-being? Indeed, valuable functioning can be very different, in relation to the conditions in which a person lives, e.g., in developing or developed areas. Some of them concern basic needs, such as being adequately nourished or being in good health, while other functioning are more complex, like achieving self-respect and being socially integrated.

Veenhoven, (2010) states that the concept of capability, despite its focus on measurable and quantitative domains, is too wide and vague, leading to lose sight of the interrelation between environmental dimensions and individual skills, due to a lack of normative and theoretical background in studies and applications (Robeyns, 2006). As an indirect answer, Sen (2003) argues that the identification and selection of relevant functioning and capabilities is an exercise that opens to doubts and dialogue, but, in defining the components of well-being, "it is undoubtedly more important to be vaguely right than to be precisely wrong".

Moreover, some theoretical issues have been raised: if capabilities are the fundamental components of well-being - and that is the reason why the authors stress the importance of freedom and choice - is a composite index, mainly built on functioning data, a reliable measure of well-being and coherent with its theoretical underpinnings?

Amartya Sen's insight and, more generally, the debate arisen about the capability approach have paved the way to the proposal of several multidimensional well-being indicators that gather a range of variables which can be connected to functioning and capabilities. Despite their limitations - in particular, the difficulties in going beyond the functioning aspects and taking into account the capabilities and their prerequisites in terms of freedom of choice, that cannot be easily measured - these indicators are interesting to our aims because in their structure the dimensions of subjective well-being begin to assume a significant role; that is why we will review some of them, in their main features, in the next section.

1.3.2   Multidimensional Well-Being Indicators

In the last decades, public and private institutions, think tanks and international organizations developed a number of well-being indicators (Fleurbaey, 2009), with different structures (both composite indicators and dashboard of indices), for several purposes and considering a great variety of dimensions. In general, the authors of the indices are aware that different users of the indicators might have different orders of priority in weighting single aspects of welfare. Therefore, in some cases, the users are allowed to choose the relative importance of each variable in the calculation of the indicator, giving the opportunity to obtain a sort of "customized" index.

What follows is a brief description of some well-known indicators, whose aim is to give basic information about their content and the methodologies applied to set them up. Some separate remarks will be devoted to the Gross National Happiness (GNH) index adopted in Buthan, due to its emblematic role: in fact, the achievement of GNH has been explicitly introduced into the Constitution of the country as a purpose of public policies. A detailed review of well-being indicators can be found in (Barrington-Leigh and Escande, 2018), where the role of subjective well-being dimensions is adequately emphasized, while Diener (2006) provides influential guidelines and recommendations for assembling national indicators of subjective well-being.

1.3.2.1 HDI: Human Development Index

The first example is directly inspired by the capability approach: since 1990 the United Nations Development Programme (UNDP) adopted the approach in elaborating its Human Development Index (HDI) and the annual Human Development Report (Robeyns, 2006). The HDI has been elaborated by Amartya Sen and the Pakistani economist Mahbub ul Haq. In measuring well-being, HDI takes into account three dimensions: health, education and material standards of living. The proxies used are: for first domain, life expectancy at birth; for the second, the mean of schooling years and the expected years of schooling; the third domain is estimated through per capita gross national income. The 2016 edition of the Human Development Report contains, in addition to HDI, four composite human development indices: the Inequality-adjusted Human Development Index (IHDI), that takes into account the variance of the HDI components within a country; the Gender Development Index (GDI), that measures disparities on the HDI by gender; the Gender Inequality Index (GII), that offers a composite measure of gender inequality using three different dimensions; the Multidimensional Poverty Index (MPI), that captures the deprivations that people in developing countries suffer in their education, health and living standards (The Human Development Report Office, 2016). Moreover, two dashboards are proposed: Dashboard 1 (Life-course gender gap), that contains a selection of indicators that measures gender gaps over the life course: childhood and youth, adulthood and older age; Dashboard 2 (Sustainable development), that contains a selection of indicators that cover environmental, economic and social sustainable development.

1.3.2.2 BLI: Better Life Index

The Better Life Index and the "How's Life" Report are the output of the Better Life Initiative launched by the Organisation for Economic Cooperation and Development (OECD) in 2011. BLI displays a dashboard of eleven dimensions "that can be considered universal" for current well-being (OECD, 2013): civic engagement, community, education, environment, health, housing, income, jobs, life satisfaction, safety and work-life balance. In the standard version of the index, these dimensions are averaged with equal weights; nevertheless, BLI is conceived as an interactive tool that allows users - through its web platform - to mix the set of proposed dimensions, giving them different weights, in order to elaborate a index coherent with their own preferences (OECD, 2017).

Within the same project, measurement of well-being is provided also on a regional basis.[6] The well-being of more than 400 regions in OECD countries is monitored along eleven components quite similar to the ones examined at the national level. As in the case of the national-based index, an interactive website allows for assigning specific weights to each variable, hence elaborating a composite well-being indicator reflecting the preference of different users and policy makers.

1.3.2.3 HPI: Happy Planet Index

Developed for the first time in 2006 by the New Economics Foundation, a British think tank, the Happy Planet Index (HPI) makes use of data on life expectancy, inequality of outcomes among people within a country, experienced well-being (the data source is the Gallup World Poll)[7] and ecological footprint (i.e., the average impact that each resident of a country places on the environment) to produce an original overview on well-being (New Economics Foundation, 2016). The aim of HPI is to give a measure of sustainable well-being; it compares how efficiently residents of different countries are using natural resources to achieve long, high well-being lives. Approximately, HPI is calculated as follows:

$$HPI \approx \frac{\text{(i.a. LE)(i.a. EWB)}}{\text{EF}}, \tag{1.1}$$

where:

i.a. LE = inequality-adjusted life expectancy;
i.a. EWB = inequality-adjusted experienced well-being;
EF = ecological footprints.

1.3.2.4 BES: Benessere Equo Sostenibile (Fair Sustainable Well-Being)

The "Benessere Equo e Sostenibile" (Fair and Sustainable Well-Being) or BES indicator (Istat, 2017) is the well-being index elaborated by the Italian Institute of Statistics (ISTAT) setting up a 12-dimension dashboard: economic well-being, education, environment, health, cultural and natural heritage, politics and institutions, research and creativeness, safety, service quality, social relations, subjective well-being, work and work-life balance. Although the conceptual and statistical similarity with BLI is evident, BES does differ from this and other examples shown above in that it avoids any sort of aggregation. The periodical reports present and discuss the entire set of proxies used to evaluate each dimension of the indicator. ISTAT does not provide any kind of

---

[6]https://www.oecdregionalwellbeing.org/
[7]The Gallup World Poll is one of the best-known surveys on behaviors and attitudes of the world population. As previously remarked, these composite indicators often include self-declared measures of subjective well-being. Surveys are the most common source for this kind of information. The crucial role of surveys in monitoring subjective well-being will be examined in depth in Sect. 1.4.

aggregation inside and between the dimensions composing BES. Conversely, the Regional Institute for Economic and Social Research of Piedmont (IRES Piemonte)[8] elaborates general and domain-specific composite BES indicators for Italian regions, presenting them in a format similar to the BLI one; as in the BLI case, the dashboard offers to the users the opportunity to assign specific weights to the components, creating an index that reflects their own priorities.

In 2018, ISTAT provided, for the first time, a system of local BES indicators for the Italian provinces. It is worth noting that, these local indices are structured on 11 dimensions only; compared to the country-aggregated BES, the subjective well-being dimension has been excluded, due to a lack of reliable information at a local level.

1.3.2.5 CIW: Canadian Index of Well-Being

Launched in 2011 (Michalos et al., 2011) with the first report on Canadian well-being, it is developed by the University of Waterloo. Similar to some previous examples, CIW is a composite single-number indicator calculated as the arithmetic mean of eight domain values: community vitality, democratic engagement, education, environment, healthy populations, leisure and culture, living standards and time use. Each domain, in turn, is composed by eight normalized indicators.

The explicit aim of the CIW proposers is to measure what GDP is unable to capture. GDP alone - they claim - cannot measure how well Canadian population is faring as a whole; GDP is not sensitive to the costs of economic growth, such as environmental degradation, and ignores the effect of growing income inequality. According to the authors, this explains why per capita GDP is growing much faster than CIW (2016).

1.3.2.6 Other Initiatives for Measuring Well-Being

In addition to the indicator previously discussed, several initiatives, mainly taken by government institutions, can be enumerated, that are on the way to producing well-being measures both complementary and alternative to GDP, or whose aim is to develop a holistic measure of national progress and well-being:

- the Australian National Development Index (ANDI)[9];

- the German W3 indicators to complement GDP (Giesselmann et al., 2013);

- the Social Report by the New Zealand Ministry of Social Development (New Zealand Ministry of Social Development, 2016);

- the study for a Global Well-being Index (GLOWING), promoted by the

---

[8]http://www.regiotrend.piemonte.it/clima-sociale/cruscotto-italia
[9]http://www.andi.org.au

University of Waterloo in the wake of CIW and adopting Kenya as a pilot-experiment country[10];

- the Measuring National Well-being programme, set up by the UK Office for National Statistics, and its interactive wheel of well-being measures[11];

- the report on well-being indicators for Israel by the Israeli Ministry of Environmental Protection (Environmental Protection, 2014).

1.3.2.7 GNH: Gross National Happiness

The Gross National Happiness Index (GNH) deserves a separate comment. GNH was introduced in the 60s by the King of Buthan Jigme Dorji Wangchuck and more recently defined by his successor Jigme Singye Wangchuck. It can be considered an original attempt to define the task of the government action in terms of individual and collective well-being; GNH, in fact, has been officially included in the Constitution of Buthan enacted in 2008, that stipulates: "The State shall strive to promote those conditions that will enable the pursuit of Gross National Happiness".

Gross National Happiness can be thought as the level of satisfaction that people in a country derive from their own life. In order to quantify GNH, a multidimensional index has been set up. The GNH index is evaluated over nine domains: community vitality; culture; ecological diversity and resilience; education; good governance; health; living standards; psychological well-being; time use. All the nine domains are equally weighted. In turn, each domain is calculated as the weighted average of several sub-indicators: on the whole, the assessment of GNH involves 33 indicators.

For each indicator, a sufficiency threshold is set. First of all, an individual happiness level is measured; to this aim, "one creates a profile for each person showing in which of the 33 indicators he or she has achieved sufficiency. Adding up the weights of the sufficient indicators gives each person a GNH score showing the share of domains in which he/she has achieved sufficiency. If a person has sufficiency in at least two-thirds, he or she is considered 'happy' in terms of the GNH index"(Buthan Studies & GNH Research, 2015).

Aggregating the individual information, the value of GNH index for the country can be obtained as the rate of happy people (H), plus the extent of sufficiency that not-happy people enjoy. This second term is calculated by multiplying the percentage of people who are not-happy (U, which is $1 - H$) by the average percentage of domains in which not-happy people have sufficient achievements (A). Therefore, the index ranges from zero to one:

$$\text{GNH Index} = H + (U \cdot A), \qquad (1.2)$$

---

[10]https://uwaterloo.ca/toward-a-global-index-of-wellbeing
[11]http://webarchive.nationalarchives.gov.uk/20160519133648/http://www.neighbourhood.statistics.gov.uk/HTMLDocs/dvc146/wrapper.html

An interesting feature of GNH is that any individual surplus over the sufficiency threshold does not contribute to a higher level of general happiness. This facet resembles the result by Kahneman and Deaton, (2010), who identified a happiness threshold around an income of $75,000; though, it should be emphasized that Kahneman and Deaton attribute this threshold to the relationship between income and emotional well-being (or happiness),[12] while GNH - despite its name - seems more related to life evaluation.[13]

1.3.2.8   Pros and Cons of Multidimensional Indicators

The development of a wide range of composite indicators of well-being is the result of the awareness - coming from the capability approach and endorsed by the Stiglitz Commission - of the inadequacy of GDP as an exhaustive welfare measure and explicitly introduces into the evaluation items referring to subjective and perceived well-being. As we have already noted, the main source of the information about subjective well-being is represented by periodical surveys on people's self-perceptions, whose attributes will be examined in Sect 1.4.

This kind of social indicators have a number of problematic features. Often, they lack solid theoretical foundations and, in many cases, they are presented without any framework for a rational discussion concerning construction (in particular, how the dimensions are weighed in the aggregation process), meaning and results. Further limitations refer to the scarce attention devoted to the correlations among the various domains and proxies; their aggregation, to create an individual well-being index, is made to the detriment of the correlations between social dimensions at individual level (Fleurbaey, 2009). This is one of the reasons why, in several cases, dashboards of indicators are provided and the weighting and aggregating step is left to the users' choice and preference.

Nevertheless, the Stiglitz Commission Report had a significant impact on well-being research, in particular - for our purposes - giving institutional acknowledgement to the relevance of subjective aspects of well-being and the usefulness of self-reported measure of it in the measurement of welfare, as well as in public and social policy-making. Apparently, in the last years, the assessment of subjective well-being in a more direct way seems to become even more feasible and reliable, due to the availability of surveys and questionnaires, and a large amount of studies have exploited these information sources in different research fields. At the same time - and mainly owing to the possible drawbacks of surveys, that will be discussed thereafter - several means and methods to evaluate welfare, starting from individual self-assessments, have been developed. In the next sections we will examine the main surveys and the methodologies applied to the evaluation of subjective well-being drawing from these data.

---

[12]See Section 1.2.

[13]In the style of GNH index, see also the Thailand national government's Green and Happiness index (Barameechai, 2007).

1.4 Self-Reported Well-Being

Different kinds of surveys are used to estimate subjective well-being and produce a wide range of indicators: worldwide surveys of a generalist nature, or specifically devoted to the evaluation of the quality of life (Gallup World Poll, World Database of Happiness, World Values Survey), generalist or quality-of-life–specific surveys on single countries, macroareas or regions (Gallup-Sharecare Well-Being Index, British Household Panel Survey, European Social Survey(ESS), Eurobarometer, Global Health & Well-Being Survey), category-specific surveys, devoted to youth (National Child Development Survey, Survey of Well-Being of Young Children (SWYC)), students, employers or employees (Social-Emotional Well-Being (SEW) Survey, GA Releases Graduate Student Happiness & Well-Being Report). In order to describe the survey approach to subjective well-being evaluation, we provide some details on the best-known surveys and the well-being indices they have generated. The main methods of well-being evaluation based on surveys or self-declared perceptions will also be discussed. Of course, both the list of surveys, indicators and methods are largely non-exhaustive.

1.4.1 Gallup Surveys

Let us start with the widely used data sources provided by Gallup, the influential analytics and advisory company surveying public opinion worldwide since 1935.

1.4.1.1 Gallup World Poll

The Gallup World Poll aims at measuring the attitudes and behaviors of the world population. It tracks a lot of issues worldwide, such as food access, employment, leadership performance and well-being. Since 2005, Gallup has conducted studies in more than 160 countries, covering - according to the company - around 99% of the world adult population. From Gallup's website[14] we learn that the survey includes more than 100 global questions as well as region-specific items; Gallup asks residents from all the monitored countries the same questions, every time, in the same way. This makes it possible to trend data from year to year and make direct country comparisons. A team of scientists advises on the development of a common set of statistics that Gallup collects in every country. The following details document how pervasive is the survey and what are the procedures followed to collect data.

Gallup uses telephone surveys in countries where telephone coverage represents at least 80% of the population or where telephone call is the common survey methodology. In the developing world, including much of Latin

---

[14]https://www.gallup.com/178667/gallup-world-poll-work.aspx

America, the former Soviet Union countries, nearly all of Asia, Middle East, and Africa, Gallup uses face-to-face interviews in randomly selected house- holds. Face-to-face interviews take approximately one hour, while telephone interviews take about 30 minutes. With some exceptions, the examined sam- ples are statistically representative of the population resident in the country and aged 15 and older. The coverage area is the entire country including rural areas, and the sampling frame represents the entire civilian adult population of the country. Exceptions include areas where the safety of the interview- ing staff is threatened and scarcely populated islands in some countries. The typical survey includes at least 1,000 individuals. In some countries, Gallup collects oversamples in major cities or areas of special interest. Additionally, in some large countries, such as China and Russia, sample sizes of at least 2,000 units are collected. In rare instances, the sample size is between 500 and 1,000. Gallup conducts the surveys on a semiannual, annual and biennial frequency that is determined on a country-by-country basis.

The questionnaire is translated into the major languages of each country and the issue of question wording and language specificities is faced through a multi-step procedure: the translation process starts with an English, French, or Spanish version, depending on the region; a translator who is proficient in the original and target languages translates the survey into the target language; a second translator reviews the language version against the original version and recommends refinements.

According to the 2008 methodological documents, the items of the survey that evaluate well-being include measures of life satisfaction, optimism, mean- ing and purpose, specific-domain satisfaction and positive and negative affect. Well-being is aggregated into five indexes: Thriving, Struggling, Suffering, Positive Experience and Negative Experience[15]:

- The Thriving, Struggling and Suffering Indexes measure respondents' per- ceptions of where they stand, now and in the future, on a scale from 0 to 10, where 0 represents the worst possible life and 10 represents the best pos- sible life. Individuals who rate their current lives a "7" or higher and their future an "8" or higher are "thriving". Those who rate their current lives as greater than "4" but less than "7" and their future lives as less than "8" and greater than "4" are "struggling". Individuals are "suffering" if they report their current or future lives as a "4" and lower.

- The Positive Experience Index is a measure of respondents' positive experi- ences on the day before the survey.

- The Negative Experience Index is a measure of respondents' negative expe- riences on the day before the survey.

The definition of the indexes resembles the distinction between the eval- uative and emotional (or experienced) components of well-being (Kahneman,

---

[15]http://www.oecd.org/sdd/43017172.pdf

et al., 1999)(see Sect. 1.1.1): in fact, the Thriving, Struggling and Suffering indexes are based on the "remembering self", i.e., life evaluation, while the Positive and Negative Experience indexes are based on the "experiencing self", i.e., capture feelings and emotions close to the subject's immediate experience.

1.4.1.2 Gallup-Sharecare and Global Well-Being Index

The Gallup-Sharecare Well-Being Index started in January 2008 as Gallup-Healthways Well-Being Index, aimed at monitoring well-being in the United States. The research and methodology underlying the well-being index is based on the World Health Organization definition of health as "not only the absence of infirmity and disease, but also a state of physical, mental, and social well-being". The well-being index measures the US residents' perceptions of life and their daily experiences through five interrelated elements that compose well-being: sense of purpose, social relationships, financial security, relationship to community and physical health.

Between 2008 and 2017 Gallup interviewed at least 500 US adults aged 18 and older daily; over 2.6 million interviews have been conducted in those years. Since it began in 2008, the Gallup-Healthways Well-Being Index survey has been conducted every day, excluding major holidays and other events, for 350 days per year. Gallup reported findings from the survey in weekly, monthly, quarterly and yearly aggregates, and by region state and community.

Since January 2018, the methodology adopted to elaborate what has become, in the meanwhile, the Gallup-Sharecare Well-Being Index underwent some changes[16]: Gallup began surveying US adults aged 18 and older using a dual mail and web-based procedure. Approximately 10,000 US adults are surveyed each month. The survey is conducted on an ongoing basis, with survey invitations sent once per month, 12 months per year.

In 2012 Gallup and Healthways created the Gallup-Healthways Global Well-Being Index to measure well-being worldwide. The Global Well-Being Index uses the same data collection and weighting methodology as the Gallup World Poll; ten questions were added to the World Poll in 2013, with each of the questions associated to one of the five components of well-being. The Global Well-Being Index is a global barometer of individuals' perceptions of well-being and is the largest recent study of its kind. The Global Well-Being Index is organized into five elements:

- Purpose: liking what you do each day and being motivated to achieve your goals;

- Social: having supportive relationships and love in your life;

- Financial: managing your economic life to reduce stress and increase security;

---

[16]https://www.gallup.com/224870/gallup-sharecare-index-work.aspx

- Community: liking where you live, feeling safe and taking pride in your community;

- Physical: having good health and enough energy to get things done daily.

The well-being level is classified according to the categories of the Gallup World Poll: "thriving", when well-being is strong and consistent in a particular element; "struggling", when well-being is moderate or inconsistent in a particular element; "suffering", when well-being is low and inconsistent in a particular element.

These are the ten statements added to the Gallup World Poll to investigate the five components of well-being. They show that the Global Index aims at measuring life evaluation and satisfaction, i.e., the cognitive component of well-being. Respondents are required to evaluate the items on a five-point scale, ranging from "strongly disagree" to "strongly agree":

Purpose:	(a)	You like what you do every day;
	(b)	You learn or do something interesting every day.
Social:	(a)	Someone in your life always encourages you to be healthy;
	(b)	Your friends and family give you positive energy every day.
Financial:	(a)	You have enough money to do everything you want to do;
	(b)	In the last seven days, you have worried about money.
Community:	(a)	The city or area where you live is a perfect place for you;
	(b)	In the last 12 months, you have received recognition for helping to improve the city or area where you live.
Physical:	(a)	In the last seven days, you have felt active and productive every day;
	(b)	Your physical health is near-perfect.

1.4.1.3 Well-Being Research Based on Gallup Data

Gallup surveys have been the data source for many studies on well-being. One of the most influential users of these surveys is the 2015 Nobel laureate Angus Deaton, who devoted a significant part of his scientific production to the analysis of well-being and - despite his long-standing familiarity with these investigation tools, or maybe because of it - has pointed out the possible fallacies of surveys in subjective well-being evaluation; these drawbacks will be examined in Sect. 1.4.5, together with the methodologies developed to circumvent them.

Studies based on Gallup data can be found, for example, in Steptoe, Deaton, and Stone (2015), where the authors present an analysis of the pattern of well-being across ages and the association between well-being and survival at older ages, and Case and Deaton, (2015), relating well-being measures to suicide rates.

Among the periodic studies on well-being that rely on these surveys, it is worth mentioning the World Happiness Report, promoted by the United Nations Sustainable Development Solutions Network[17] and released almost yearly since 2012. The report ranks 156 countries according to the well-being perceived by their citizens. The ranking follows a Cantril ladder method, that basically implies that respondents rate their own well-being perception on a 0 to 10 scale (Levin and Currie, 2014). The main data source of the report is the Gallup World Poll, integrated occasionally with other sources, such as the World Values Survey (see Sect. 1.4.3).

Another research initiative - which is interesting to our aims - has been launched in 2020 by the collaboration between Gallup and the European Commission.[18] Around 360,000 people in 117 countries have been classified based on whether they live in a city, a town or a rural area, and have been asked to rate their own lives on a 0 to 10 scale. The study shows that people living in urban areas report, on average, a higher level of life satisfaction. The substantial "urban *vs* rural" gap is not related to the country income level. The surveyed countries include United States and countries where the Gallup World Poll is conducted through face-to-face interviews; this clearly means that the developing areas of the world are overrepresented. Nevertheless, the research highlights the importance of having locally disaggregated data available in well-being evaluation. This, together with the high frequency of the data, is one of the advantages of the well-being indicator we propose in this book.

1.4.2 European Social Survey

The European Social Survey (ESS) is an academically driven cross-national survey that has been conducted, since 2002, every two years across Europe, through face-to-face interviews with newly selected, cross-sectional samples.[19] The survey measures the attitudes, beliefs and behavior patterns of populations in more than 30 nations. The ESS data are available for non-commercial use and can be downloaded from the website. The stated aims of the surveys are:

- to chart stability and change in social structure, conditions and attitudes in Europe and to interpret how Europe's social, political and moral fabric is changing;

- to achieve and spread higher standards of rigors in cross-national research

---

[17]https://https://worldhappiness.report
[18]https://news.gallup.com/opinion/gallup/315857/degree-urbanisation-effect-happiness.aspx
[19]https://www.europeansocialsurvey.org

in the social sciences, including for example, questionnaire design and pre-testing, sampling, data collection, reduction of bias and the reliability of questions;

- to introduce soundly based indicators of national progress, based on citizens' perceptions and judgements of key aspects of their societies;

- to undertake and facilitate the training of European social researchers in comparative quantitative measurement and analysis;

- to improve the visibility and outreach of data on social change among academics, policy makers and the wider public.

A key aim of the ESS is to improve methodological standards in the field of cross-national surveys. As we will see in Chapters 3 and 5, comparability is a crucial issue in well-being evaluation, both when the measurement is based on objective quantities - that often turn out only apparently comparable - and *a fortiori* when self-perceived well-being is involved.

In order to achieve "optimal comparability" across countries, an ESS scientific team produces a detailed project specification, which is revised in light of each successive round. National teams of participating countries should read the specification in its entirety to ensure that fieldwork is conducted according to the same standards cross-nationally. This "principle of equality or equivalence" applies to sample selection, translation of the questionnaire, and to all methods and processes associated with data collection and processing.

The ESS has been collecting methodologically robust cross-national data on well-being since 2002. The survey includes headline measures of subjective well-being such as "life satisfaction" and "happiness" as part of its core questionnaire, submitted to respondents in each round. More in-depth data on well-being are also provided by some rounds, where thematic "rotating modules" have focused on different aspects of well-being. These data on well-being are collected along with a large number of socio-demographic background variables and questions about social and political topics, providing a rich dataset.

A module that focuses on the "personal and social well-being" of respondents was first introduced in ESS3 (2006), and then repeated in ESS6 (2012). The ESS6 module also sought to incorporate a new validated scale of positive well-being, and included questions to develop the evidence base on behaviors that promote well-being. While the core ESS items measuring subjective well-being ("life satisfaction" and "happiness") mainly follow the hedonic well-being definition which emphasizes positive feelings, the rotating module on "personal and social well-being" focuses on the eudemonic approach and evaluates several aspects such as autonomy or self-determination, interest and engagement, positive relationships, sense of meaning and purpose in life. The module also links personal well-being to social well-being, emphasizing the importance of interpersonal and societal-level experiences and behaviors.

A module on "family, work and well-being" was included in ESS2 (2004) and ESS5 (2010), exploring the inter-relations between work, family and well-being in a comparative perspective. The repeat of the ESS2 (2004) module in ESS5 (2010) also enables to study consequences of economic recession on employees' well-being. The module includes several questions measuring general well-being as well as questions measuring job satisfaction and satisfaction with work-life-balance.

A module focusing on "social determinants of health and health inequalities" was first fielded in ESS7 (2014). Specific items included a range of health measurements (Body Mass Index [BMI], self-reported diagnoses and mental well-being), social determinants (childhood conditions, housing quality and working environment), behaviors (smoking, alcohol use, fruit and vegetable consumption and physical activity), and use of primary, secondary and alternative health care.

Moreover, a depression scale has been introduced in rotating modules on "personal and social well-being" and "social determinants of health and health inequalities". The scale consists of eight items which ask about positive and negative emotions; the items can either be analyzed as a set to measure depression or single items can be linked to other sub-concepts and analyzed separately.

It is worth noting the initiative "Measuring and Reporting on Europeans Well-Being: Findings from the European Social Survey" which showcases the scope that ESS data provide for exploring the definition, distribution and drivers of subjective well-being across Europe, and encourages academics, policy makers and students to use the ESS website resource for carrying out their own research and informing policy choices.[20]

Fors and Kulin (2016) is an example of application of the ESS data source to the cross-country study of subjective well-being. The authors construct new measurement instruments that capture both the affective and cognitive dimensions of subjective well-being. Using ESS data and multi-group confirmatory factor analysis, they estimate latent country means for the two dimensions and compare country rankings across the two measures. The results reveal significant differences in country rankings depending on whether one focuses on affective well-being or life satisfaction. Quite important to our aims, the authors point out that life satisfaction is more susceptible to social norms and other biases, and hence suggest that the affective component deserves more attention in comparative cross-national research; this aspect will be recalled in Chapters 3 and 5, where our subjective well-being indicator will be applied to different countries and cultural contexts.

---

[20]https://www.europeansocialsurvey.org/docs/findings/ESS1-6_measuring_and_reporting_on_europeans_wellbeing.pdf

1.4.3 World Values Survey

The World Values Survey[21] (WVS) is managed by an international network of social scientists studying changing values and their impact on social and political life. The survey was designed to test the hypothesis that economic and technological changes are transforming the basic values and motivations of the population of industrialized societies. The WVS, which started in 1981 focusing on developed societies, nowadays consists of nationally representative surveys conducted in more than 100 countries which contain around 90% of the world's population, using a common questionnaire.

The WVS is a cross-national and time-series investigation of human beliefs and values, currently including interviews with almost 400,000 respondents, and cover a wide range of situations, from very poor to very rich countries, in different cultural areas. The survey measures and monitors topics such as support for democracy, tolerance of foreigners and ethnic minorities, support for gender equality, the role of religion and changing levels of religiosity, the impact of globalization, attitudes toward the environment, work, family, politics, national identity, culture, diversity, insecurity and subjective well-being.

The freely available dataset has been a source for several studies. For example, Sun et al. (2016) analyzes how subjective well-being in a Chinese population varies with subjective health status, age, sex, region and socio-economic characteristics.

1.4.4 European Quality of Life Survey

The European Quality of Life Surveys (EQLS)[22] is carried out every four years since 2003. The survey examines both the objective features of European citizens' lives - employment, income, education, housing, family, health and work-life balance - and subjective topics, such as people's levels of happiness, how satisfied they are with their lives and how they perceive the quality of their societies.

The geographical coverage of the survey has expanded over time, as a consequence of the prospective European enlargements. The last wave is extended to the 28 EU members and to five candidate countries: Albania, FYR Macedonia, Montenegro, Serbia and Turkey.

The EQLS has developed into a set of indicators which includes environmental and social aspects of progress and complements traditional indicators of economic growth and living standard such as GDP or income. The data collected have made possible to track key trends in the quality of European citizens' lives over time. For example, the report "Quality of life in Europe: Trends 2003-2012"[23] compares the results from the first three waves of the

---

[21]http://www.worldvaluessurvey.org
[22]https://www.eurofound.europa.eu/surveys/european-quality-of-life-surveys
[23]https://www.eurofound.europa.eu/publications/report/2014/
quality-of-life-social-policies/quality-of-life-in-europe-trends-2003-2012

survey to provide evidence of trends and change in the quality of life of citizens in European countries over a decade. It also examines whether differences across EU members have narrowed or remained stable; one of its findings is that subjective well-being has remained stable across the EU as a whole, but it also finds that financial strain in households has grown in the wake of the economic crisis.

The study by Soukiazis and Ramos (2016) represents an example of analysis of subjective well-being conducted by means of EQLS data. The authors analyze the determinants of life satisfaction (as the evaluative component of subjective well-being) and happiness (as the emotional component) of the Portuguese citizens using data from the European Quality of Life Survey.

1.4.5 How to Collect (and Interpret) Self-Reported Evaluations

Self-reported evaluations are extensively used to study subjective well-being. Nevertheless, a sizeable group of studies casts shadows on the reliability of self-declared measures of well-being. A quite obvious point is that surveys usually require a large amount of resources; therefore they turn out to be expensive and cannot be carried out with a high frequency.

More specifically, self-reported well-being evaluations are considered sources of potentially biased information. In particular, these studies document that reports of well-being may be influenced by manipulations of current mood and of the immediate context, that often cannot be avoided in surveys. (Schwarz, 1999, Schwarz and Strack, 1999) emphasize that a self-evaluation is conditioned by the information that is most accessible at a point in time: when the interviewer asks "taking all things together, how would you say things are these days?", the respondent usually limits his/her own evaluation to the most recent and frequently recalled experiences. If this is the case, preceding questions in a survey may bring information to mind that respondents would otherwise not consider. Interesting examples are provided by Deaton (2012) and deepened in Deaton and Stone (2016); in brief, the authors find that both evaluation of life as a whole and the emotional well-being prove to be sensitive to question order effects. In particular, asking political questions before the life evaluation questions significantly reduces reported life evaluation: people may dislike politics and politicians so much that prompting them to think about them has a remarkable downward effect on their assessment of their own lives. In some cases, these order effects persist deep into the interview, and condition both the reporting of hedonic experience and of satisfaction with standard of living.

On the whole, the analyses based on self-reported well-being have shown several vulnerabilities (Kahneman, Krueger, et al., 2004):

- Specific circumstances have unexpected small effects on well-being.

  The reason seems to be the propensity of individuals to adapt to circumstances: positive and negative events (marriages, widowhood, physical deseases, extraordinary increases in income such as lottery wins) exhaust their effect on well-being in a relatively short time.

Different interpretations of the evidence are proposed: people progressively adapt to positive and negative situations, thus "absorbing" the effect on well-being (so called "hedonic treadmill"); positive and negative affects may persist over time, but their evaluation is relative to expectations, and expectations adjust to circumstances, thus reducing the impact of events on perceived well-being (so called "aspiration treadmill").

An example can be found in Deaton (2012), where the author uses Gallup data on well-being to analyze the reaction of US residents to the first years of the 2008 economic crisis. The study shows that, although in the early days of the crisis the respondents reported lower levels of life satisfaction and greater anxiety, these measures had largely recovered by the end of 2010, despite the fact that high levels of unemployment indicated that the crisis was ongoing. It follows that subjective well-being measures seem to do a much better job of monitoring short-run levels of anxiety than they do of reflecting the evolution of the economy over a year or two; in other words, the measures reflect the emotional component of well-being better than its evaluative part and are not good proxies of the economic performance of a community.

- Seemingly similar countries show unexpected and large differences in life satisfaction.

This reveals that self-evaluations of well-being may reflect the differences in cultural and social norms that determine self-perception.

- People seem unable to provide unbiased evaluations of experiences extended over time.

The reason is that individuals assign little or no weight to the duration of experiences, in evaluating their own quality of life. Therefore, the duration of positive or negative events or feelings have limited role in life evaluation, that requires retrospective assessment.

In order to overcome these potential biases, Kahneman, Krueger, et al. (2004) suggest that measures of subjective well-being should meet the following requirements:

1. they should represent actual hedonic and emotional experiences as directly as possible;

2. they should assign appropriate weight to the duration of different segments of life;

3. they should be minimally influenced by context and by standards of comparison.

The same authors discuss the methods implemented in survey-based research to collect information about personal feelings and experiences. The methods are intended to reduce some of the bias sources previously defined:

- The *Experience Sampling Method* (ESM) - which is similar, in its aim, to the Ecological Momentary Assessment (Walz, Nauta, and Rot, 2014) - is carried out providing the respondents with an electronic diary that beeps at random times during a day and asks them to describe what they were doing just before the prompt. Respondents are also required to register the intensity of their feelings. The procedure should overcome the problems raised by direct questions about global life satisfaction; in particular, imperfect recall and duration neglect. Unfortunately ESM is not a practical method for national well-being accounts, mainly because it is impractical to implement in large samples and because infrequent activities are only rarely sampled. Moreover, the interruption of normal activities is also thought as a remarkable disadvantage of the methodology (Kahneman, Krueger, et al., 2004).

- The *Daily Reconstruction Method* (DRM) is more practical; it asks respondents to keep a diary corresponding to events of their previous day. Besides being required to describe their feelings, respondents indicate, for each event, when the event began and ended; what they were doing; where they were; whom they were with. All this should help to achieve an accurate recall, despite the retrospective approach of the procedure.

- In the same line of DRM, the *Event Recall Method* (ERM) is easier to administer, particularly in telephone surveys. ERM asks questions about feelings associated with particular events in the recent past. Empirically - according to Kahneman, Krueger, et al. (2004) - DRM and ERM produce similar results, even if the selection of specific events may yield differences in self-reported well-being when respondents have heterogeneous preferences in their time allocation choices.

An additional issue in collecting self-evaluations on well-being through surveys is that individuals may interpret and use the response categories differently. To circumvent the problem, survey researchers try to anchor response categories to words that have a common and clear meaning across respondents, but there is no guarantee that respondents use the scales comparably. Therefore, one could legitimately question whether the numeric values attached to individuals' responses about their life satisfaction or emotional states should be given a cardinal interpretation, because of the potential personal use of scales. As a possible solution, Kahneman and Krueger (2006) propose the U-index, an indicator which - building on the information gathered with the methods previously described - measures the proportion of time that people spend in an unpleasant state; the index has the virtue of being an ordinal measure (of "poorly being") and not requiring a cardinal conception of individuals' feelings.

In conclusion, surveys' users have to face the challenge of constructing measures or indicators that allow for disentangling different components of well-being: a component due to the affect generated by a given set of events, and a component related to the allocation of time across different situations. Intuitively, this is strictly related to the usual challenge of well-being estimation - that we have previously discussed - i.e., separating emotional aspects of well-being from life evaluation.

Furthermore, the issue of interpersonal comparison should not be neglected; in fact, satisfaction with life and with particular domains is affected by comparisons with other people and with past experiences (Clark, 2003).

Finally, a problem of intercultural and international comparison has been mentioned; significant differences in self-reporting of well-being can be noted - both in life satisfaction and in affective perception - due to the influence of social norms and cultural habits, sometimes filtered by language expressions.

Although the accumulated knowledge about the process underlying self-reports, in recent years, have improved questionnaire design, several issues are still open, that deserve the research of alternative or complementary information sources for the evaluation of subjective well-being. The next section is devoted to examining the opportunities provided by the large-scale datasets coming from social network sites. Not all the issues raised in this section will find an exhaustive solution and, not infrequently, new questions will be prompted. Nevertheless - thanks both to the availability of a huge amount of high-frequency data and to innovative investigation methods - interesting perspectives will be unfolded about the evaluation of subjective well-being.

1.5   Social Networking Sites and Well-Being

The remarks about survey-based research that we have discussed in the previous section seem to emphasize that what social scientists would like to do, when they investigate subjective well-being, is *to ask* individuals for a self-evaluation of their own well-being status *without asking* them for anything, i.e., avoiding to condition their answer and, if possible, letting them ignore they are observed. This, in fact, reminds of the "Hawthorne effect" or "observer effect", that is the phenomenon by which individuals modify their behavior in response to their awareness of being observed (Landsberger, 1958); it is a dilemma analysts are quite familiar with, even if it has been argued (Di Tella and MacCulloch, 2008) that social scientists (and economists in particular) typically measure what people do rather than listen to what people say.

Things have rapidly changed with the advent of information technology, the internet and social networking sites (SNS); the increasing frequency of social interactions through digital devices have provided scholars and observers of social phenomena with new large-scale dataset. SNS, in fact, host an enormous amount of records that can be collected and analyzed for research purposes, making it possible to study social dynamics from an unseen perspective, minimizing the interaction between the researcher and the observed individuals (Pentland, 2014). The simultaneous availability of these large-scale sources of potential information and the increase in computational power of technological structures may have created, initially, some disorientation and, sometimes, a misleading overconfidence in the possibility to know and control individual and collective behaviour. Fortunately, thanks to significant progress in statistical theory and applications, social sciences are fostering their capacity to manage and analyze dataset with large dimension and high capillarity (King, 2011, Lazer et al., 2009), extracting meaning and separating information from background noise. It should be noted, moreover, that ethical issues have been raised about a fair use of the information provided by SNS.

1.5.1 Sentiment Analysis

Sentiment or opinion analysis is the core aspect of a new investigation method for monitoring public opinion that can be applied to the measurement of well-being. This research field is largely dedicated to the systematic extraction of web users' emotional state from the texts they post autonomously on different internet platforms, such as blogs, forums, social networking sites (e.g., Twitter or Facebook) (Kramer, 2010, Ceron, Curini, and Iacus, 2013, Curini, Iacus, and Canova, 2015). Social psychologists have found a link between well-being of individuals and their use of language; in other terms, it is possible to extract words from SNS messages, allowing to reconstruct the emotional content the author wanted to communicate, infer psychological traits and measure well-being (Quercia, 2015).

The availability of large-scale datasets, in particular from the Internet, has driven up the growth of a significant number of theories and methodologies for sentiment analysis. The traditional approach to sentiment analysis mainly focuses on the volume of data, trying to capture people's attitudes simply counting words or tabulating frequency of mentions. These methods - known as *unsupervised* and largely automated - are often based on the use of ontological dictionaries: a text is assigned to a specific opinion category if some predetermined words or expressions do (or do not) appear in the text (Grimmer and Stewart, 2013, Ceron, Curini, and Iacus, 2015). The benefit of the approach is the opportunity to implement an automated analysis, once the dictionary has been defined.

On the contrary, more advanced methods - known as *supervised* - aim at a more intrusive approach: the largest part of the classification work is automated, but a crucial role is still assigned to the intervention of human coders. The methods examine the structure of the language and can appreciate nuances of meaning, informal speech, jargons, paradoxical or ironic expressions, that often are missed or misinterpreted by a totally automated analysis (Hopkins and King, 2010).

Despite its limitations (Couper, 2013), if correctly performed, methods of sentiment analysis seem to be a useful framework to evaluate subjective well-being, when the drawbacks of standard survey methodology appear too severe (Iacus, 2014, King, 2016). On one hand, in fact, there is no need for asking questions to the investigated population; all the analyst has to do is *to listen* to the online conversations and classify the opinions expressed accordingly. On the other hand, the available information is updated in real time (and sometimes the updating comes for free); hence the frequency of the well-being time series can be as high as desired.

As argued in the previous sections, one of the main issues in evaluating subjective well-being is the need to separate the emotional (or short run) from the evaluative (or medium-long run) parts. A high frequency indicator of well-being gives the opportunity - through the statistical analysis of stochastic processes - to untangle a *structural component* of the phenomenon from its *volatility*, following their evolution over time; assuming that evaluative well-being is more stable than emotional well-being, structural and volatile components of the well-being time series appear as good proxies of the two well-being dimensions.

What we present in the following chapters of the book is a new supervised method of sentiment analysis (in Section 2.7), based on a text-mining algorithm that extracts meaning from conversations on SNS. An easy-to-interpret indicator of subjective well-being will be obtained (see Section 3.5) and its potential will be shown through some applications in Chapters 3–5; in particular, we discuss how the indicator can integrate official measurements of well-being (Chapter 4) and how international and intercultural comparisons (Chapters 3 and 5) may be encouraged by a method that is, basically, language independent. Chapter 5 will present an extensive analysis of what the different SNS-based well-being indexes report abound the 2020's COVID-19 pandemic.

Before that, let us rapidly review the wide set of studies we can find in the literature, attempting to estimate subjective well-being using data from SNS.

1.5.2   Evaluating Subjective Well-Being on the Web

Studies aiming at tracking well-being through social network communication have been surveyed by Luhmann (2017) and Scollon (2018), which we refer to for an exhaustive classification. The advantage of social network sites as data sources for the evaluation of well-being is usually found in the abundance of so-called "digital traces" left by web users, that are available on a large scale, at low cost, without the explicit consent of their authors, and that are updated in real time. Some of these claims are, to some extent, extreme, but it is undeniable that websites such as Twitter, Facebook and Instagram or services like Google Trends provide a huge amount of potentially new information to the assessors of social sentiment.

Different kinds of data have been analyzed to obtain an estimate of well-being: frequency of posting, number of Facebook "likes" and even smiling on profile pictures (Lee, Efstratiou, and Bai, 2016, Kosinski, Stillwell, and Graepel, 2013, Seder and Oishi, 2012), but mostly the digital traces used are, of course, the texts posted by the web surfers.

Based on the language patterns adopted by web users, Luhmann (2017) identifies two main methodologies: a *closed vocabulary* and an *open vocabulary* approach. The first one classifies the texts according to keywords coming from a predefined lexicon. The best-known example of closed dictionary for subjective well-being evaluation is the Linguistic Inquiry and Word Count (LIWC)[24] which, in its 2015 version, is composed of almost 6,400 words, word stems and selected emoticons. LIWC reads a given text and counts the percentage of words that - according to the dictionary - reflect different emotions, thinking styles or social concerns. Each dictionary word is classified into one or more categories, that are arranged hierarchically, indicating positive or negative sentiment or specific emotions; when a word is found in a text, the scores of these categories is incremented. On the contrary, the "open vocabulary" is a bottom-up (or data-driven) approach: the list of words is not predefined, and is rather inferred from an examination of the text, that identifies relevant words, stems or linguistic features (see, e.g., Schwartz, Sap, et al. (2016)). As pointed out by Kern et al. (2016), a "closed vocabulary" approach makes it easy to count word frequencies, but it ignores that language can be ambiguous, particularly in SNS conversations: punctuation, unconventional terms and even misspellings or mistypes may reflect feelings and emotions that cannot be captured by a predefined list of words and stems. On the other hand, "open vocabulary" methods require a larger amount of data and more sophisticated statistical tools to read texts and interpret results. Combinations of closed and open vocabularies are not excluded (Mikolov, Chen, et al., 2013a, Turian, Ratinov, and Bengio, 2010).

The studies aim at the measurement of *individual* or *aggregated* well-being. The different measurement level is mainly due to differences in the algorithms implemented in the analysis[25]; when a sufficient number of observations are available for the same web user, the technology allows for an evaluation of the individual subjective well-being. This approach, however, raises the issue of the aggregation criterion to adopt, when the researcher needs to infer something about social well-being from individual evaluations; that is the reason why, regardless of the availability of individual information, some procedures directly estimate the proportion of positive/negative feelings in the aggregated text, thereby providing an evaluation of social well-being. According to Grimmer and Stewart (2013), shifting focus from individual level to estimating proportions can lead to substantial improvements in accuracy, even if the examined text set is not statistically representative of the totality of potential expressions (Hopkins and King, 2010). Interesting intermediate

---

[24]http://liwc.wpengine.com/
[25]See Grimmer and Stewart (2013) for details.

aggregation levels are represented by local or geographic evaluations, when data are geo-localized.

For example, the algorithm presented in this book allows for an estimate of the aggregate subjective well-being. Nevertheless, as far as geo-localized texts are used, a local-level breakdown of the estimate will be available.

To some extent related to the aggregation level are ecological and exception fallacies. *Ecological fallacies* (King, Rosen, and Tanner, 2004) imply drawing conclusions about phenomena estimated at one level, based on data collected at another level. An example by Gelman et al. (2008): on one hand, at the individual level, wealth is positively correlated to tendency to vote Republican; on the other hand, wealthier states tend to vote Democratic. Beyond the possible explanations provided by the authors - e.g., voting preference may be determined by self-perceived relative wealth more than by wealth in absolute terms - what we know is that aggregate-level correlation does not necessarily coincide with individual-level correlation. Similarly, *exception fallacies* can also occur, when conclusions about groups are made based on outliers or exceptional cases; Kern et al. (2016) recommend a careful and discretional human supervision and the adoption of estimation models robust to outliers.

Scollon (2018) surveyed the most recent studies on well-being based on social network sites information, classifying them according to the data source and pointing out, along with their added value, their possible limitations:

Facebook. A *Gross National Happiness*[26] has been launched by Facebook in 2009, with the aim of measuring collective well-being and tracking it over time, classifying the status updates posted by users. The methodology does not seem dissimilar from the "closed vocabulary" approach; in fact, as a member of the data team wrote, Facebook "adapted a collection of positive and negative emotion words built by social psychologists. Examples of positive or happy words include 'happy', 'yay' and 'awesome', while negative, or unhappy words, include 'sad', 'doubt' and 'tragic'. At the same time, the company guaranteed anonymity and respect of privacy, ensuring that "no one at Facebook actually reads the status updates in the process of doing this research; instead, our computers do the word counting after all personally identifiable information has been removed"; so, the method appears totally unsupervised.

Nevertheless, some researchers had the opportunity to correlate the index scores to daily life satisfaction of Facebook users, in order to test its reliability. Kramer (2010) examines the correlation between the Facebook Gross National Happiness Index and life satisfaction expressed by a sample of over 1,000 Facebook users; the estimated coefficient is positive, but low in absolute terms, even if correlation shows an increase in conjunction with events that are particularly positive or negative. On the contrary, Wang,

---

[26]There are many indexes called like this: the one of Section 1.3.2.7 developed in Buthan, this Facebook index and the Twitter-based index by Rossouw and Greyling (2020).

Kosinski, et al. (2014) find a non-significant correlation (moreover, with a negative coefficient) between the two measures and conclude that Facebook Gross National Happiness Index is not a valid well-being indicator. Scollon (2018), however, notices that Facebook users that provided the life satisfaction scores do not necessarily belong to the users' set examined to evaluate Facebook Gross National Happiness Index; this may contribute to explain the result. Liu et al. (2015) offer an alternative explanation: they assume that social norms induce Facebook users to post more positive status updates than their real well-being status would suggest, while negative expressions are less biased self-evaluation of well-being; therefore, positive status updates would be less correlated to actual subjective well-being than negative emotions are. These results raise an endogeneity issue in well-being estimation via social network sites, that will be briefly discussed in the following; in particular, they suggest that SNS users may be biased when posting their own status updates, quite similarly to what happens to survey respondents (see Section 1.4.5). The problem, in fact, is relevant also for the well-being indicator we propose in this book. As a crucial difference, however, our indicator does not rely on self-declared status of internet surfers but only on the language they use in their conversations; in our opinion, this reduces significantly the probability to register an intentionally biased well-being evaluation. Moreover, the authors find that only recent events (i.e., only more recent status updates) have an impact on subjective well-being, which is another trait this literature has in common with the results we have previously examined about surveys.

Twitter. Since the indicator we propose in this book is based on Twitter messages, particular attention is devoted to studies on this site. The platform, with more than 300 million users, is a huge source of texts, that nevertheless is not free from quantitative and qualitative issues, which have been pointed out, among the others, by Salvatore, Biffignandi, and Bianchi (2020). Due to the brevity of the messages allowed, Twitter is often assumed to be more suitable to evaluate emotional well-being than life satisfaction (Luhmann, 2017).

Happiness. For example, Dodds, Harris, et al. (2011) build a happiness indicator, called "Hedonometer", based on the frequency of use of a set of words for which they have obtained happiness evaluations on a nine-point scale, using Mechanical Turk.[27] Their dataset is huge, being made of around 4.6 billion expressions posted by over 63 million Twitter users in the period from September 2008 to September 2011. The authors analyze the temporal patterns of happiness, in particular weekly and daily cycles, finding subjective well-being peaks in the week ends and between 5 and 6 a.m. The same authors apply the Hedonometer to study assortative phenomena of

---

[27]Mechanical Turk is a crowdsourcing website owned by Amazon, providing services connected to internet, such as collection of data and categorization of products or images. See https://www.mturk.com/

Twitter users with respect to well-being (Bliss et al., 2011). Applications of the Hedonometer can also be found in Frank et al. (2013) while a technical description of the method is presented in Section 3.2 as well as some applications in Section 5.3.

The Hedonometer is now evaluated daily by the University of Vermont Complex Systems Center, which can provide, therefore, a time series since 2008.[28] The indicator has also been applied to geo-tagged data subsamples, in order to obtain evaluations of well-being in specific regions.

Several studies exploit the opportunity offered by the Hedonometer to geo-localize tweets and examine local dynamics of well-being. Mitchell et al. (2013) explore a dataset composed by over 10 million geotagged tweets gathered from 373 urban areas in the United States in 2011. The authors elaborate classifications of states and cities based on their similarities in word usage and connect word choice and message length with urban features such as education levels and obesity rates. Further, they estimate the happiness levels of states and cities, correlating them with demographic data; as a result, happiness within the United States is found to correlate strongly with wealth, showing large positive correlation with household income and strong negative correlation with poverty and obesity. Not surprisingly, a significant driver of the happiness score for individual cities is found to be frequency of swear words.

As we already noted in the Facebook case, the success of the Gross National Happiness, introduced in the 60s in Buthan as an alternative measure of collective well-being (see Section 1.3.2.7), has generated a subjective well-being index with the same name, based on Twitter posts and evaluated since 2019 in three Commonwealth member countries: South Africa, New Zealand and Australia.[29] The aim of the project is to measure, in real time, the sentiment of countries' citizens during different economic, social and political events (Rossouw and Greyling, 2020, Greyling, Rossouw, and Adhikari, 2020).

In order to calculate the *Twitter Gross National Happiness index*, sentiment analysis is applied to a live Twitter feed, and each tweet is assigned either a positive, neutral or negative sentiment. Then, an algorithm evaluates a happiness score on a 0-to-10 scale. The Gross National Happiness index provides a happiness score per hour for each of the three countries but now expanding to several other countries. More technical details can be found in Section 3.3 as well as some applications in Section 5.2.

The approach adopted by the Gross National Happiness index is, in many respects, similar to the methodology followed by the Subjective Well-Being index, proposed and discussed in this book. In both cases, in fact, the sentiment is drawn from a reading and classification of tweets, that involves the work of human coders, while the evaluation of collective well-being is

---

[28]https://hedonometer.org/timeseries/en_all/
[29]http://gnh.today

entrusted to an algorithm. Nevertheless, the evaluation method of the Gross National Happiness index still relies, to some extent, on the meaning of the words used in the texts to express a feeling, particularly when the analysis goes deep into the breakdown of sentiment into specific positive or negative feelings (Rossouw and Greyling, 2020). On the contrary, SWB and the iSA algorithm make use of all the terms (not only the words with a particular meaning) and the language structure to infer mood, hence exploiting suggestions coming also from jargon, neologisms, paradoxical or ironic expressions.

Greyling, Rossouw, and Adhikari (2020) apply the Gross National Happiness index to the study of happines in South Africa during the COVID-19 pandemic. They show a significant reduction of happiness, due to the lockdown and its aftermaths in terms of isolation, depression, anxiety and uncertainty about the future. More interestingly, the authors evaluate the probability to be happy with and without the lockdown restrictions - also estimating, in this second hypothesis, the increase in the number of COVID-19 cases - and draw the conclusion that people in South Africa would have been happier with a higher spread of the virus and no lockdown regulations than vice versa (see Section 5.2).

Finally, an interesting "happiness paradox" is pointed out by Bollen, Gonçalves, Leemput, et al. (2017). They examine a set of about 40,000 Twitter users and estimate individual levels of subjective well-being, collecting the 3,200 most recently tweets submitted by each user and performing a sentiment analysis based on the OpinionFinder[30] subjectivity lexicon. The authors discover that the majority of social media users are less happy than their friends are on average. One of their conclusions is that social media use may be associated with increased levels of social dissatisfaction if individuals are subject to unfavorable comparison between their own happiness and popularity to those of their friends. A similar procedure is followed in Bollen, Gonçalves, Ruan, et al. (2011), studying assortative dynamics in social networks.

Life satisfaction. Despite the view that tweets can only contribute to measure short-run emotional well-being, Schwartz, Eichstaedt, et al. (2013) study the language used in tweets in about 1,300 different US counties, with at least 30,000 tweets per county and correlate the subjective well-being expressed by Twitter users to life satisfaction measures coming from surveys available for each county, using an approach based on lexica and topics (clusters of lexico-semantically related words). The authors find that Twitter language use improves accuracy in predicting life satisfaction over and above standard demographic and socio-economic variables included as controls (age, gender, ethnicity, income and education). One of the main limitations of the study is - as in the case of Wang, Kosinski, et al. (2014) - that individuals surveyed and individuals whose tweets are analyzed may differ.

Other studies make use of Twitter data to monitor life satisfaction. For example, Yang and Srinivasan (2016) conduct a passive surveillance of life

---

[30]https://mpqa.cs.pitt.edu/opinionfinder/

satisfaction expressions on Twitter, applying an original methodology; starting from the Satisfaction With Life Scale (SWLS) designed in 1985 by Diener, Emmons, et al. (1985), they retrieve the tweets conveying self-ratings of life satisfaction from a two-year collection of Twitter data (about 3 billion tweets) and evaluate individual well-being from both a static and a dynamic perspective. Their main result is that, unlike emotional well-being, trends in life satisfaction are immune to external events (political, seasonal). The method allows for interpersonal comparison of life evaluation (the authors point out differences between satisfied and dissatisfied users in several linguistic and psychosocial features) and for observing evolution of life satisfaction over time (they notice, for example, that the psychosocial tweet features of users who change their status from satisfied to dissatisfied or vice versa are quite different from those who stay in the same status over time). On the other hand, Scollon (2018) emphasizes that the methodology is still based on pre-defined lexica and that researchers need to carefully select their words for searches when trying to extract well-being from tweets. Finally, it should be noted that, overall, the number of life satisfaction tweets was very low, compared to the total number of tweets. On the one hand, this seems an implicit support to the opinion that Twitter better reveals the affective and emotional components of well-being than the structural ones; on the other hand, it revels a fallacy of the method, that is crucially based on the analysis of tweets that contain an explicit evaluation of life satisfaction. On the contrary, the method we propose in this book estimates subjective well-being making use of the language of all the tweets posted, regardless of their specific content.

Other examples of well-being analysis based on Twitter data can be found in (Lim et al., 2018, Durahim and Coşkun, 2015, Abdullah et al., 2015, Quercia et al., 2012, Greco and Polli, 2020).

It should be noted that the methodology adopted to extract sentiment and meaning from the tweets is not neutral with respect to the results. Jaidka et al. (2020), for example, exploit geo-localization of tweets to estimate subjective well-being at the county-level in the United States. They make use of both closed-dictionary (e.g., LIWC) and data-driven methods. In order to check the robustness of the estimates, they use the Gallup-Sharecare Well-Being index as a reference target and find that closed-dictionary methods can yield estimates of county well-being inversely correlated with Gallup survey estimates, due to regional cultural and socioeconomic differences in language use, while estimates seem to be more robust when supervised data-driven methods are applied. The result emphasizes the potential role of cultural norms in affecting the terms used to express feelings, and denotes - what is most important to our aims - the inadequacy of word-level methodologies in interpreting the huge amount of information provided by social media; the study, in other words, supports the use of supervised procedures, such as the iSA algorithm proposed in this book.

Weibo. Sina Weibo is a Chinese microblogging site - a sort of Chinese version of Twitter - that boasts more than 500 million registered users. Using large-scale dataset from this platform, Zhao et al. (2018) examine the relationship between subjective well-being and some macroeconomic indicators in Chinese cities in 2016. The study applies a user-defined dictionary approach and find an inverted U-shaped relationship between city residents' well-being and GDP per capita: in poor cities, the well-being increases when GDP per capita increases; in cities where per capita income is higher, on the contrary, perceived well-being decreases as GDP per capita further increases. A similar relationship is estimated between city residents' well-being and population size: well-being increases with population sizes in small cities, but decreases in cities where the population size exceeds a certain threshold. On the other hand, the authors find a positive relationship in all cities between well-being and the available share of green space.

Beyond SNS? The same criticism made to surveys has been, to some extent, raised to the investigation of well-being via SNS posts. A group of studies, in fact, observe that the use of social media may cause a distortion in self-perception and expression of well-being; when people are aware of sharing their opinions with a community, a self-presentation incentive may lead to well-being reports that do not properly reflect true feelings and emotions (Ruths and Pfeffer, 2014). As an alternative data source, these studies suggest to use internet searches (in particular, Google search activity) - which are conducted in privacy - as "one of the rare glimpses into private behaviors that is available to scientists" (Scollon, 2018).

Ford et al. (2018), for example, explore the potential of internet search data to serve as indicators of subjective well-being. Their hypothesis is that searches for positive and negative affect-related terms represent information-seeking behavior of individuals who are experiencing emotions and seeking information about them. Data on the frequency of Google searches for 15 affect terms have been collected from Google's Trends website and correlated with data on health, self-reported emotions, psychological well-being, personality, and Twitter postings at the state and metro area levels in the United States. They find that internet search scores correlate with indicators of cardiovascular health and depression; in particular, affect searches predict depression rates at the metro area level beyond the effects of income and other well-being measures. Moreover, some search term scores also correlate strongly with self-reported emotions, well-being metrics, neuroticism, per capita income and Twitter postings. It should be observed, however, that the positive correlation of Google searches and both subjective well-being (measured by Gallup-Healthways survey) and affect-related Twitter posts implicitly reinforces the belief that the three sources are measuring the same phenomenon in similar ways. Moreover, the study is significantly confined to search of negative affect terms; in fact, one of the main limitations of the data source is that people experiencing negative feelings are more likely to seek - even on the net - ways to face them; people with higher well-being levels, on the contrary, are less likely to spend time in Google search of positive emotion terms. This means that not even Google searches can be considered unbiased with respect to the emotional status. Aware of these limitations, the indicator we propose in this book tries to exploit the language used in all the SNS conversations, irrespective of being affect-related or not, in hopes to reduce the incidence of these sources of bias.

An interesting example is offered by Algan et al. (2016), who propose an indicator of individual subjective well-being in the United States based on Google Trends data from January 2008 to January 2014. The authors retrieve from Google Trends groups of keywords whose behavior over time fits with the weekly time series of survey-based well-being measures provided by Gallup Analytics. Moreover, a model is built to predict the well-being measures obtained through surveys; the good out-of-sample performance of the model allows for suggesting the use of web-based indicators to complement survey-based or official statistics, exploiting the advantages of web data in terms of high frequency and geo-localization. The study also identifies the keyword categories (job search, financial security, family life, leisure) that better predict well-being variations. It is worth noting that this indicator has much in common with the subjective well-being index presented in this book; in fact, both the indicators aim at improving official statistics, providing high frequency and local estimates of well-being, usually not available at an official level. On the other hand, they differ in that Algan et al. (2016)'s indicator relies on Gallup Analytics surveys to select the keywords that can capture the day-by-day evolution of well-being, while our index - as a sentiment analysis instrument - tries to extract the feeling directly from the language used by the web surfers.

1.5.3   Pros and Cons of Large-Scale Data from SNS

Nowcasting. Large-scale data coming from the web seem particularly suitable to a monitoring activity known as *nowcasting*. The term is a contraction for "now" and "forecasting" and is quite common in meteorology: it refers to the opportunity to collect information about some quantities or variables in real time.

Two main situations make nowcasting interesting for social well-being analysts. First, several variables related to well-being - such as unemployment rate and GDP - are observable, but only with a significant time lag; consequently, traditional measures reflect well-being changes with a delay and do not provide policy makers with a prompt information. As previously pointed out, therefore, SNS data can aim at complementing low-frequency official statistics.[31] Secondly, important natural and social events that affect well-being - such as earthquakes, terrorist attacks and acts of violence in

---

[31]Interestingly, Benjamin et al. (2020) agree with the opportunity of complementing

schools - are unpredictable; therefore, any impact analysis on well-being has to rely on data collected at a significant time distance from the event. On the contrary, conversations on social networks - on Twitter, in particular - allow for studying real-time changes in emotional well-being as a reaction to unexpected and extraordinary occurrences (Miura et al., 2015, Jones et al., 2016).

Endogeneity. A potential endogeneity issue has been raised about monitoring well-being through social network activity. May self-perceived or self-declared well-being be biased by the use of social networks itself? (Hall, 2016, 2017), for example, in two different experiments, explored the relationship between happiness and internet use on a sample of college students, applying an internet use scale. The experiments reveal that people who report spending less time on the internet, less time expressing emotions and more time checking facts score higher on measures of happiness.

The topic is, in fact, more general and transversal, concerning not only evaluations grounded on SNS data, but also survey-based methods. Do survey modes affect the content of self-declared well-being? Interestingly, Schork, Riillo, and Neumayr (2019) show that the impact of survey mode depends on the content of a survey and its variables. In particular, objective variables (e.g., employment status) are not affected by a mode-specific measurement bias, while subjective variables (e.g., job satisfaction) are; in their study, web respondent report lower satisfaction levels on subjective variables, compared to telephone respondents.

The relationship between internet use and emotional status is also investigated by experiments where web use is exploited to alter subjective well-being; mobile phone applications monitor users' emotions, cognitive and motivational states, activities, environmental and social context, and predict their mood, based on previous well-being self-report. The devices also provide users with specific advice on how to improve their mood (Burns et al., 2011).

Ethical concerns. Somewhat related to possible endogeneity of SNS monitoring results are the ethical concerns about studies on individual and social sentiment based on internet data. Generally, the main criticism emphasizes that social network users are unaware they are providing research data when they post a text on the net, and hence they cannot give any explicit consent. More critical objections have been raised about a few experiments where information via SNS has been manipulated to condition subjective well-being perception and measure web users' reactions (Kramer, Guillory, and Hancock, 2014, Panger, 2016).

It must be emphasized, in this regard, that all the texts we use for feeding

---

official statistics with self-reported well-being measures, but are critical of survey-based indicators.

our indicator are acquired via the public Twitter API. In other words - even if it remains true that no express permission has been provided by the authors of the messages - Twitter users are aware that their conversations are public and can be read by anyone on the net. On the other hand, no question is asked to Twitter users and no interference is created that can modify their mood and the content of their messages.

Representativeness. A representativeness point, on the other hand, sheds some doubts about usefulness and reliability of SNS data, arguing that users of these platforms are not a representative sample of the whole population. For example, older people are usually underrepresented; according to surveys on the topic, Instagram is more appealing for young people, female and urban residents; Pinterest users are more concentrated in the female 25–34 years old class. Several studies suggest to solve the problem - which is, to some extent, in common with online surveys (Grewenig et al., 2018) - adjusting the data according to the differences in demographic features between SNS users and the entire population, thus obtaining an estimate of the public opinion (Kern et al., 2016).

It should be conceded, however, that representativeness has become a problem also for traditional surveys, either in face-to-face or phone or web mode. In fact, in addition to the progressive decrease in the diffusion of house phones, it can be noted that people are increasingly reluctant to take part into surveys, due to the high frequency of requests they receive from different communication channels (e-mail, house and mobile phone, SNS).

Most of all, the purpose of capturing the dynamics of public opinion makes the statistical representativeness of surveyed samples less important; public sentiment is, in fact, volatile and hence surveys only allow for a snapshot of a situation that rapidly changes. Describing a dynamic effect would require a long series of surveys, that are too expensive, both in terms of human and economic resources.

Moreover, a survey based on a representative sample basically aims at capturing the average mood of the population. This is actually not the only interesting dimension of public opinion in current sociological research. The information coming from SNS is significant - despite its representativeness limitations - as far as it can be considered the source of a sort of "lighthouse effect"; it is a quite common (and debated) view, in fact, that SNS anticipate and, in several cases, affect public opinion and social trends. The issue is relevant in forecasting analyses and in studies where the variable of interest is the opinion trend more than the average opinion. In this perspective, most of the representativeness problems are overcome; what is significant, in fact, is not the statistical and demographic composition of internet users' population, but rather the volume or, at most, the percentage of texts expressing a sentiment. For example, "retweets" - which are often considered a source of bias, because they seem to alter artificially the relative importance of an opinion - are not a matter of concern, because they simply indicate that internet surfers will meet these feelings with a higher frequency and will be more probably affected by them.

Reliability. Lastly, a common objection to the use of SNS data to estimate social sentiment is the observation that web users may hide their true identity and give false information; in other terms, the web is a great source of data, which unfortunately are unreliable because the internet is "full of liars". The remark is well argued, but it should be noted that:

- estimates of well-being obtained by the indicator proposed in this book are not affected by this kind of lies. The indicator never uses the identity of Twitter users or the news they convey in their tweets: it simply takes into account the mood expressed through the language, which is independent of the reliability of the statements. That is why, for instance, the issue of fake news is ruled out from our analysis.

- it is hard to understand why a Twitter user should lie more frequently than a person required to answer to a survey. In this case, in fact, potential anonymity would be an advantage, making it possible for everyone to express an opinion without revealing his/her own personal identity and data. On the contrary - as it has been previously documented - strategic lies are quite frequent in surveys, just because people are aware they are observed by an interviewer.

It is not clear, in other words, why the reliability of information drawn from SNS data should be *a priori* questioned, compared to alternative sources that exhibit the supposed advantage of a higher level of representativeness.

1.5.4 International and Intercultural Comparisons

Several studies examine the language specificities observed in SNS communication (Baym, 2015, Omori, 2014). Some of them also analyze the strategic use of language to create relational networks via web (Zappavigna, 2012). Less attention seems to be paid to problems raised by the comparison of opinion and sentiment expressed in different languages and cultural contexts. As an example, a recent study by Pew Research Center provides a detailed and interesting tableau of religious commitment among people living in European countries[32]; the evolution of the overall religiosity is evaluated by a composite index measuring the importance assigned to religion in one's life, regularity in religious service attendance, prayer frequency and belief in God. Comparison issues may arise if we assume that the propensity to reveal and express one's religious convictions differs from one context to another, according to cultural reasons or social norms. Further problems may occur if different languages use different terms to express similar feelings, showing an involuntary heterogeneity in the intensity of opinions and beliefs.

---

[32]https://pewrsr.ch/2rjepvr

Some of the methods of sentiment analysis - and the *iSA* algorithm we use in this book is among them - provide an opportunity to monitor different linguistic and cultural environments at the same time, due to their language-independence and to the availability of large dataset coming - basically with no significant difference in access barriers - from corners of the world that are far apart. For this reason the issue of inter-linguistic and inter-cultural comparability of sentiment (particularly, in the case of self-evaluations) is assuming an increasing relevance in web-based social research with respect to traditional survey-based studies.

On the other side, the availability of SNS conversations gives, to some extent, an unprecedented opportunity to let these differences emerge, both related to cultural habits and language use, and to take them into account in subsequent research. An example will be provided in Chapter 3, where a comparison of the SWB indicator will be carried out between two different social contexts such as Italy and Japan.

1.6 Subjective or Social Well-Being?

As we told in the Introduction, a possible misunderstanding may arise in identifying the exact content of our analysis: are we trying to measure subjective well-being in an individual or in a collective perspective? The answer is conditioned, in general, by the measurement method adopted; when someone is asked for providing an evaluation of his/her own well-being status, and even when we use SNS conversations as a source, we are supposed to obtain information about *individual* well-being, i.e., subjective well-being *stricto sensu*. This would leave a problem unsolved: how should individual well-being be aggregated, in order to get an evaluation of *collective* (or *social*) well-being? The issue is interesting both from a theoretical and an empirical viewpoint, and is not dissimilar from the problem of aggregation of individual preferences in microeconomic analysis of social welfare.

Far from entering the complex debate - whose technical implications will be discussed in the next chapter - we simply point out that some procedure exist that, starting from individual data, allow for directly measuring the collective phenomenon, without evaluating individual well-being; one of these methods, that circumvent the aggregation issue, is *iSA*, the algorithm described and applied in this book.

1.7 Glossary

Capability approach: The capability approach to well-being measurement is based on a view of living as a mix of beings and doings, where the quality of life can be measured in terms of valuable functioning and capabilities.

Eudaimonic well-being: A sense of well-being derived from living according to one's values, being one's true self, reaching one's full potential and developing personally. It is related to sense of purpose and meaning in life.

Evaluative well-being: It captures overall life satisfaction or fulfillment, as global judgments about life as a whole.

GDP: Gross domestic product.

Hedonic well-being: Sometimes defined as "experienced" well-being. It is centered on positive emotions, pleasure, satisfaction. Feelings such as sadness, anger, illness and pain diminish this well-being perception.

SNS: Social networking sites.

Well-being: Status of good health and good fortune, or more generally, related to wellness.

2

Text and Sentiment Analysis

2.1 Text Analysis

Text analysis, sometimes referred to as text mining, consists of extracting information (semantic content) from textual data. While for a human it is relatively easy to perform such task, automatic methods grasp in performing well in all situations. On the other side, humans fail when it comes to analyze huge amounts of texts. In this book we distinguish the term "sentiment", when it is related to the sign, e.g., positive/negative, from the term "opinion" (why positive? why negative?).

The term opinion mining has been introduced for the first time by Dave, Lawrence, and Pennock (2003) to indicate a technique capable of conducting a research on a given set of keywords and to identify attributes for each term (positive, neutral, negative), so as to, once the statistical distribution of these three terms is available it would be possible to extract the opinion associated to each keyword. Almost all the subsequent works focused on producing essentially sentiment classification linked to individual keywords (Pang, Lee, and Vaithyanathan, 2002, Pang and Lee, 2004). The term sentiment analysis has also replaced the term text mining which actually refers to textual analysis in a broad sense (Liu, 2006).

The main research question related to this book could be *"How do people feel about …?"* As already discussed in Chapter 1, the Internet and the vastness of content in SNS in particular, are inexhaustible sources of textual data containing valuable information. But even before the outburst of the WWW and social media, linguists along with statisticians and computer scientists, have adapted well known techniques and developed new ones in order to extract the sentiment and the opinions from digital texts. As in every field of research, each technique has its pros and cons and in fact there is no "best technique" or a universal one, although it is still possible to identify reasonable and efficient techniques for classes of problems and types of data sources.

One of the most common wrong beliefs about textual (and social media) analysis is that methods which rely on a brute force approach, like letting computers to automatically extract metrics, counting how many times a certain word or a hashtag appears, counting the number of followers of a Twitter account, counting the number of likes of a post on Facebook, and so forth, do work. Despite the fact that those are in fact useful information to a certain extent, for instance they could be useful to study the structure of a given personal/group network of relationships and its strengths, they do not produce any useful information in terms of semantic content.[1]

2.1.1   Main Principles of Text Analysis

As already mentioned above, by opinion analysis we mean the ability to extract the reasons behind a positive or negative sentiment.

The technique called iSA (integrated Sentiment Analysis) is, as explained in Section 2.7, a technique of estimation of the distribution of opinions providing good statistical properties which does not rely on computer science alone (Ceron, Curini, and Iacus,, 2016). But just as no body escapes Newton's laws, no technique can escape the fundamental principles of text analysis (Grimmer and Stewart, 2013):

- **Principle 1.** *Every quantitative linguistic model is wrong, but some can be useful.* The mental process that leads to the production of a text, without exception, is purely a mystery. Even for the finest linguists or psychologist. Every single sentence can drastically change its meaning due to the inclusion or exclusion of very small bits of language. The following text is an authentic review of a film on the DB Moive Review portal. It reads: *"This film has **good** premises. Looks like it has a **nice** plot, an **exceptional** cast, **first class** actors and Stallone gives his **best**. But it **sucks**"* The last three words (and especially the term "but") completely change the meaning of that statement in the previous sentence. In fact, although there is a predominance of positive terms (five positive against one negative term) that three-word sentence, completely switches the semantic content of the whole text. Most of the time, it is the use of the so-called functional words (articles, nouns, prepositions, etc.) that characterize a sentence although these functional terms alone do not mean anything and are usually discarded in automated sentiment extraction methods (Pennebaker, Boyd, et al., 2015). This is typical in Twitter data, where the use of a hashtag, somehow ironically, at the end of a comment can completely change the interpretation of the tweet. It is also the motivation why, in many situations, the "bag of words" approach seems more useful than the structured one. Oftentimes it is the positioning or the absence of punctuation that can distort the meaning of a text, but in typical text analysis, punctuation is usually removed. Think about the famous prophecy attributed to Sibyl Latin: *"Ibis redibis numquam peribis in bello"*, which can be translated as *"will go, will come back, will not die in war"*, but also the opposite way, *"will go, will not come back, will die in war"*. Word jokes, metaphoric or ironic sentences like: *"there is no*

---

[1]As an example, if we were simply focused on counting the number of followers of the two contenders for the White House in the presidential election of 2012, Obama with about 16 million followers, would have won 94 to 6 (with a gap of 88 percentage points) against the candidate Romney who had just one million followers at the beginning of the campaign. The difference between the two was finally less than 4%.

favourable wind for the mariner who doesn't know where to go" (Seneca), double meaning terms, etc., make natural language interpretation a great fun for humans and a painful job for computers.

- **Principle 2.** *Quantitative methods help, but cannot replace human.* For the reasons mentioned above the automatic methods can only make some operations over texts faster and allow to scale up to large corpora. They can be considered a tool that enhances or aids human capabilities (such as a telescope or a lever), but they are certainly not a tool to replace the human brain.

- **Principle 3.** *There exists not best or ideal technique of text analysis.* Each technique is designed with very specific purposes and is based on assumptions defined *a priori*. In addition, in the case of text analysis, there are additional constraints, such as the language itself (Italian, English, etc.), the topic of discussion (politics, economy, sports), the historical period (the same words can be hot or cold depending on the historical period), the age and gender of the writer and the nature of the interlocutors (imagine a text in which two students, or a student and a teacher or two teachers discuss about exams), and so forth. Moreover, there are also techniques designed for **individual classification** of texts and others designed for **aggregated classification**. For individual classification algorithms the aim is to attribute a semantic category (or an author, a topic, etc.) to each un-read text. Whereas in aggregated classification, the object of interest is to study the aggregate distribution of semantic categories (or opinions, topics, etc.) over the population of texts. In short, rather than search for the proverbial needle in the haystack (individual classification), we try to figure out the form the haystack takes (aggregated classification). In social science, opinion and sentiment analysis are intrinsically linked to the concept of aggregated classification, especially in forecasting tasks. In fact, it is usually more important to know who wins the elections rather than who each elector voted for. One could argue that the distinction between individual and aggregate classification is artificial, since the individual classification, after all, can also be used to produce the aggregate distribution. Unfortunately, the ex-post aggregation is highly dangerous, if not detrimental, as will be discussed in Section 2.6.

- **Principle 4.** *Validation of the analysis.* Every new method, as well as every model, must be validated by the data itself. The *supervised* methods, i.e., those for which the semantic categories are known *a priori* or are identified by manual coding on a subset of texts called the training set, can be easily validated in every analysis by cross-validation, especially if we consider the individual classification techniques. This validation can be performed by checking the semantic classification generated by the method and the objective (or hand-coded) semantic meaning of each texts. For *unsupervised* methods, where the semantic categories are identified *a posteriori* through

the observation of recurrences within groups of texts classified as homogeneous or the assignment is made by cross-checking dictionaries of terms or
catalogs, the validation is a particularly difficult task, if not impossible on
a large scale. In such circumstances, the analysis may require the construction of controlled experiments, such as entering text for which you know
the semantic content, but of which the algorithm ignores the classification,
and verifying that the method assigns the document to the group that it is
assumed to be correct.

2.2   Different Types of Estimation and Targets

As mentioned in the previous section there are two large families of methods in
data analysis, and textual analysis is just a particular type of data analysis.
One family of methods is referred to as **supervised** and the other **unsupervised**. For simplicity, let us denote by Y an outcome variable (e.g., the
sentiment) and by $X = (X_1, X_2, \ldots, X_K)$ a vector of K features/covariates
(e.g., words). Most machine learning methods are also supervised methods
because they require a **training set** and a **test set**. A *training set* is a subset
of the data for which the response variable Y is known in advance or it has
been learned through human supervision (hand coding). This set of data for
which the variable of interest is known ex ante, is used to train the machine
learning algorithm $A(\cdot)$ of the type

$$Y^{train} \sim A(X^{train}) + \epsilon,$$

where we denoted by (Y^{train}, X^{train}) the values of the outcome variable and
feature vectors for the training set. Here the symbol $\sim$ is used in the statistical
sense to mean *"distributed as"* and ϵ is a statistical error.[2] Then, the trained
algorithm is applied to the *test set*, which is the remaining set of data for
which the outcome variable has to be estimated (or predicted), i.e.,

$$\hat{Y}_i = A_{trained}(X_i^{test}), \quad i \quad \text{belongs to test set}$$

where $\hat{Y}_j$ is the predicted value for a new data j with feature vector $X_i^{test} =
(X_{i1}, X_{i2}, \ldots, X_{iK})$. For example, one can take as training set a certain num-
ber of tweets, hand code them according to the sentiment (positive, neutral,
negative and off-topic), train an algorithm and apply the trained algorithm
to new tweets in the pipeline.

   Most machine learning methods perform **individual** classification (or estimation), which means, given a new data X_i, the algorithm provides an estimated value of the response variable Y_i accordingly. Very few algorithms

---

[2]Like in simple regression analysis for which the above formula becomes $Y \sim a + b'X + \epsilon$.

focus instead on the **aggregated** classification, i.e., the estimation of the final distribution of possible values of the outcome variable (e.g., opinions) $P(Y)$. In the social sciences, oftentimes **target** of estimation is exactly $P(Y)$ (i.e., which is the distribution of positive, negative, neutral? Which is the distribution of votes after and election?) rather than individual classification (which tweet is positive? Who will vote for which party?). Of course, once individual classification is available, it is still possible to aggregate estimated values to obtain the final distribution. While this is acceptable in well-designed experiments, i.e., when random sampling from a given distribution is available, it is not most of the times in social media or observational studies in general. We will discuss in detail why this happens in Section 2.6, while Table 2.1 summarizes this discussion.

For unsupervised methods, the target can be the topic of discussion or the position on a latent axis (e.g., left-right or positive-negative) or the position of each sentence in a suitable space of words. These methods can also have supervised counterparts. We will discuss briefly each method in what follows, although our real interest in well-being estimation is the aggregated distribution estimation presented in Section 2.6.

One additional comment, before reviewing a few traditional machine learning methods commonly used in textual analysis, is the evaluation of the error ϵ. Clearly, each statistical method is designed to reduce the error ϵ according to some loss or cost measure.[3] This is where the fourth principle in Section 2.1.1 matters when a method of analysis has to be adopted.

2.3 From Texts to Numbers: How Computers Crunch Documents

One essential step in textual analysis is transforming a set of documents into machine-digestible forms. We call **corpus** a set of documents consisting of texts. A collection of corpus is called **corpora**. The natural way to transform texts into numbers is to split each document in words and prepare a matrix in which each row represents a document and each column represents a word. The element in row i and column j contains the number of times word j has been used in document i. This matrix is called **Document-Term** matrix. This matrix is potentially very large due to its number of columns,[4] even if the number of documents is very small and usually the set of documents in social media analysis is very large. This approach is called **bag-of-words**, as the order of the words is disregarded. The words are no longer called as such but rather **tokens**. But, despite the complexity of human language, most of

---

[3]For example, in linear regression, also known as least squares method, the loss function $\sum_i (\hat{Y}_i - Y_i)^2$ is the one to be minimized.

[4]Think of English language: the Oxford Dictionary includes about 650,000 terms.

Table 2.1
Classification of methods with example of candidates. *Note.* NLP = Natural Language Processing; SVM = Support Vector Machines; LLS = Local leader stage.

	Type of learning algorithm	
	Unsupervised	**Supervised**
Target of estimation		
Individual estimation	Corpora approach, NLP, WordFish, topic models, word2vec, clustering methods, etc.	SVM, Random Forests, Artificial Neural Network, Deep Learning, WordScores, LLS, etc.
Aggregated estimation	Aggregation of the above	ReadMe, iSA

the information can be retained by dropping a large part of unnecessary terms (columns). This step is called pre-processing and consists of **lowercasing** all texts (i.e., Room, ROOM and room are collapsed to room), removing **stop words** (which are language dependent, but normally consist of prepositions, articles, conjunctions, etc.), remove **numbers**, removing **punctuation** and also removing words which are **too frequent** (in a discussion about "sex", the word "sex" itself exists for all texts and therefore is not informative) or **too rare** (a word that is used very few times and in very few texts is not statistically representative of the set of words used in the corpus), etc.

Tokens later pass a process of **stemmization** and/or **lemmatization**. Stemming algorithms work by cutting off the end or the beginning of the word, taking into account a list of common prefixes and suffixes that can be found in an inflected word. The idea is to reduce a set of words to their fundamental common root, e.g.: *family, families,* can be unified into the stem *"famil"* but *familiar* should not; although this unification may occur in practice. Lemmatization, on the other hand, takes into consideration the morphological analysis of the words. To do so, it is necessary to have detailed dictionaries which the algorithm can look through to link the form back to its lemma. For example, while in stemming the two tokens "studi-es" and "study-ing", after removing the suffixes, have two different stems "studi" and "study", in lemmatization they correspond to the same lemma "study", as "studies" can be interpreted as the third person, singular number, present tense of the verb study. These two steps are not prone to error as they depend on the language and content.

The order of words and more complex words can be sometimes important. Clearly, if rude stemming or lemmization is applied to the sentence. *"The White House in Washington D.C"* the two tokens "white" and "house" are treated as in the sentence *"I own a white and small house by the sea"*. Stems (or tokens or lemmas) prepared in this way are called *unigrams*. In the above case it is better to consider the *bigram* "white_house" instead. As one can imagine, triads of words are called *trigrams*, and so forth.

Generally speaking, considering stems with three or more words does not provide an addition of information and does not increase the quality of the classification apart from the specific examples given above and in fact, the most used stemming procedures are limited to unigrams.

One remark is due for Asian languages. While for Western languages the stemming is usually performed looking at blank characters (space) between words, in most Asian languages, like Japanese or Chinese, there is no space between words. Words may be a sequence or two or three characters and sometimes a mix of alphabets (like in Japanese). So a preliminary step before pre-processing, consists of the **tokenization** of texts to which the pre-processing is then applied.

Further, the counts of the token in each document, does not necessarily add information to the analysis compared to a simple presence/absence indicator. Therefore, after the pre-processing the final document-term matrix remains still large but treatable by computers. There is still an additional step to understand. Think about this silly dialog:

"What is it?"
"That is a dolphine."
"No, it is a killer whale!"

The above can be transformed, after removing punctuation, into a document-term matrix like this:

Document/Term	what	is	it	that	a	dolphine	no	killer	whale
Document 1	1	1	1	0	0	0	0	0	0
Document 2	0	1	0	1	1	1	0	0	0
Document 3	0	1	1	0	1	0	1	1	1

The characteristic of the above matrix is its **sparsity** which means that, over $3 \times 9 = 27$ elements, only 13 out of 27, i.e. 48.1%, are not zero and hence with a sparsity of 51.9%. With very large corpus, along with the growth of the matrix also, and usually, the sparsity may get larger and larger. This is a good feature of the document-term matrix because there exist memory efficient methods for storing sparse matrixes but also an issue when there is the need of using it in linear algebra calculations, which most traditional methods rely on.

An additional empirical fact to know is that, in any particular topic of discussion, the number of words used, after preprocessing, amounts to a number between 200 and 500.

2.3.1 Modeling the Data Coming for Social Networks

Although we present the algorithms in their very general form, it is important
to model the data in the frame of Web crawling or Social Media. Let us denote
by $\mathcal{D} = \{D_0, D_2, \ldots, D_M\}$ the set of possibile categories (i.e., sentiments or
opinions) where D_0 is the category assumed to be the most relevant in the
data and absorbing most of the probability mass of $\{P(D), D \in \mathcal{D}\}$, i.e.,
the distribution of opinions in the corpus of N texts. From now on, this is
outcome variable Y and denoted as D unless otherwise specified. Normally
(but see below) D_0 refers to the texts corresponding to **Off-Topic** or not-
relevant texts with respect to the analysis (i.e., the *noise* in this framework)
but which are commonly present in any crawled set of texts from the social
network.[5] Let S_i, $i = 1, \ldots, K$, be a unique vector of L possible stems (i.e.,
single words, unigrams, bigrams, etc. which remain in the document-term
matrix after the pre-processing phase) which identifies one of the texts in
a corpus. For example, in document 1 *"What is it?"*, the vector of stems
s_1 is $s_1 = (1, 1, 1, 0, 0, 0, 0, 0, 0)$, for document 2 *"That is a dolphine"*, $s_2 =
(0, 0, 0, 1, 1, 1, 0, 0, 0)$ and for document 3 *"No, it is a killer whale!"* we have
$s_3 = (0, 1, 1, 0, 1, 0, 1, 1, 1)$. In practice a stem vector is a row of the document-
term matrix. In this trivial example, all stem vectors are different and hence
the unique stem vectors are $S_1 = s_1$, $S_2 = s_2$ and $S_3 = s_3$, but in a real
corpus, more than one text can be usually represented by the same unique
vector of stems S_i; this is why we discriminate between unique stem vectors
S_i and document (individual) stem vectors s_i.

Each vector S_i belongs to $\mathcal{S} = \{0, 1\}^L$, the space of 0/1 vectors of length
L (the number of stems), where each element of S_i is either 1 if that stem is
contained in a text, or 0 in case of absence. The number of possible different
rows in the document term matrix is then $K = 2^L$.

In a corpus of N texts, we have s_j, $j = 1, \ldots, N$, vectors of stems associated
to the an individual text/document j and s_j can be one and only one of the
possible $S_i, i = 1, \ldots, K$. As $\mathcal{S}$ is, potentially, an incredibly large set (e.g., if
$L = 10$, $2^L = 1024$ but is $L = 100$ then 2^L is of order 10^{30}), we denote by $\bar{\mathcal{S}}$ the
subset of $\mathcal{S}$ which is actually observed[6] in a given corpus of texts and we set
$\bar{K}$ equal to the cardinality of $\bar{\mathcal{S}}$. To summarize, the relations of the different
dimensions are as follows: $M << L < \bar{K} < N$, where "$<<$" means "much
smaller". In practice, M is usually in the order of 10 or less distinct categories,
L is in the order of hundreds, $\bar{K}$ is in the order of thousands and N can be up
to millions. The data set is then formalized as the set $\{(s_j, d_j), j = 1, \ldots, N\}$
where $s_j \in \bar{\mathcal{S}}$ and d_j can either be "NA" (not available or missing) or one of
the hand-coded (or tagged using a trustable algorithm) categories $D \in \mathcal{D}$. We

---

[5]For example, in a TV political debate, any non-electoral mention to the candidates or
parties are considered as D_0, or any neutral comment or news about some fact, or pure
Off-Topic texts like spamming, advertising, jokes, etc.

[6]As mentioned, the usual dimension after pre-processing is around 200–500.

denote by $\Sigma = [s_j, j \in N]$ the $N \times \bar{K}$ matrix of stem vectors of the whole corpus.

The typical aim of the analysis is the estimation of aggregated distribution of opinions $\{P(D), D \in \mathcal{D}\}$ from a corpus of texts by using doing individual classification of each single text in the corpus, i.e., predict $\hat{d}_j$ from s_j, and then tabulate the distribution of $\hat{d}_j$ to estimate $P(D)$.

We assume that the subset of tagged texts is of size $n << N$. In this set there is no misspecification. This will be the *training set*. The remaining set of texts of size $N - n$, for which $d_j = \text{NA}$, is the *test set*.

2.4   Review of Unsupervised Methods

2.4.1   Scoring Methods: Wordfish, Wordscores and LLS

The idea of scoring consists of ordering texts along a spectrum of continuous opinions rather than classifying, them according to a discrete and finite set of categories. Scoring techniques refer to a more general framework called Item Response Theory (IRT) (Boeck and Wilson, (2004)) originated in psychometry and psychology. This theory assumes the existence of a latent dimension, which is the fictitious axes on which text lie. A typical application of these scoring techniques in political analysis is that of ordering electoral speeches or those documents written by political actors, e.g., the scoring along the left-right axis. These techniques can be implemented in both supervised or unsupervised approach. Among the unsupervised techniques, we can mention **Wordfish** (Slapin and Proksch, 2008). This technique produces an ordering based on the frequency of terms contained in the texts but does not provide any clue on the meaning of the underlying axis. Indeed, once the ordering is available, the outcome should be analyzed further by looking at the texts which have been classified at extremes of the axis and infer from a qualitative analysis on them, possible polarities like positive/negative, left/right, materialism/post-materialism, etc.

The unit of analysis is the "author" i or, better, one of his/her texts. Wordfish analyzes the word frequencies within a text under a Poissonian model. For each text i, the frequency y_{ij} of a word j is assumed to have a Poisson distribution with constant rate λ_{ij}, i.e.,

$$y_{ij} \sim Poisson(\lambda_{ij})$$

with

$$\lambda_{ij} = \exp\left(\alpha_i + \psi_j + \beta_j \cdot \omega_i\right)$$

where α_i is the fixed effect for document i, ψ_j is the fixed effect for word j and β_j is a specific weight/importance of word j in discriminating documents.

Finally, ω_i is the position of document i along the latent dimension. The algorithm is numerically intensive as it uses the EM (Expectation-Maximixation) algorithm and it consists of the following steps: the E-step calculates the expectation of latent variable and the M-step maximizes the log-likelihood estimated in the E-step. More precisely,

- Step 1: initial value for $\psi_j^0 = $ log of mean count of each word j; α_i^0 relative log-ratio of mean word count of each doc i with respect to the first document in the corpus. Single Value Decomposition is used to obtain initial values of β_j^0 and ω_j^0.

- Step 2: the log-likelihood: for each document i, the following likelihood is maximized:

$$\ell(\alpha_i, \omega_i) = \sum_{j=1}^{m} (-\lambda_{ij} + \log(\lambda_{ij}) \cdot y_{ij})$$

where $\lambda_{ij} = \exp(\alpha_i + \psi_j^0 + \beta_j^0 \cdot \omega_i)$;

- Step 3: the log-likelihood for each unique word j

$$\ell(\psi_j, \beta_j) = \sum_{i=1}^{n} (-\lambda_{ij} + \log(\lambda_{ij}) \cdot y_{ij}) - \frac{\beta_j^2}{2}\sigma^2$$

is maximized, where $\lambda_{ij} = \exp(\alpha_i^{step2} + \psi_j + \beta_j \cdot \omega_i^{step2})$, assuming a prior distribution for $\beta_j \sim N(0, \sigma^2)$.

- Step 4: calculate the global log-likelihood

$$\sum_{j}^{m} \sum_{i=1}^{n} (-\lambda_{ij} + \log(\lambda_{ij}) \cdot y_{ij})$$

- Iterate Steps 2–4 until the gain in log-likelihood is negligible.

This approach has several limitations though. For example, if texts are too small (like tweets), the method is unstable and convergence is not granted. Moreover, if the chosen value of σ (which is a tuning parameter) is too small, each single word can split two documents apart. Finally, it is computationally intensive and the results do not allow for an intuitive interpretation of the latent variable as the method is completely unsupervised. Despite these limitations, the method is still valuable as it is not dictionary based and hence works with texts written in any language.

One of the most famous supervised scoring algorithm is called **Wordscores** (Laver, Benoit, and Garry, 2003). This method assumes that there exists a source of already classified texts along one or more dimensions and therefore produces a scoring along these predefined axes (e.g., left to right, positive to negative, etc.). Wordscores essentially implement simple analyses of frequencies as follows.

For each reference text $r = 1, \ldots, R$, we assume to know its position on the axes corresponding to dimension d. Its a-priori/known position is denoted by A_{rd}. Let F_{wr} be the relative frequency of each different word w used within text r. Doing this for each word w and reference text r produces a matrix of relative frequencies from which it is possibile to calculate a new matrix of conditional probabilities. It is possible to construct the probability to read from reference text r given that word w is read:

$$P_{wr} = \frac{F_{wr}}{\sum_r F_{wr}}.$$

Suppose we have text A and B. The word "Japan" appears 10 times per 10,000 words in A and 30 times per 10,000 words in B. Then, with probability 0.25 we are reading from text A, and 0.75 from text B (assuming this word only appears in A and B). We can build the score for each word as follows:

$$S_{wd} = \sum_r (P_{wr} \cdot A_{rd}).$$

Notice that P_{wr} and A_{rd} are given a priori. If reference r is the only text containing word w, then $P_{wr} = 1$ and $S_{wd} = A_{rd}$.

To continue the example, if text A has position -1 and Text B has position $+1$, then the word "Japan" has the following score:

$$0.25 \cdot (-1) + 0.75 \cdot (+1) = +0.5.$$

Suppose to have a set of unlabeled texts $u = 1, \ldots, U$. Denote by F_{wu} the relative frequency of word w in the unlabeled text u. Then, the score of text u along dimension d is

$$S_{ud} = \sum_w (F_{wu} \cdot S_{wd}).$$

The predicted score of text u is affected by commonly used words with average score

$$S_{ud} = \sum_w (F_{wu} \cdot S_{wd}).$$

To avoid this problem, scores S_{ud} are transformed so that they have a similar dispersion of the reference texts in the following way:

$$S_{ud}^* = (S_{ud} - \bar{S}_{ud}) \left(\frac{\sigma_{rd}}{\sigma_{ud}} \right) + \bar{S}_{ud},$$

where $\bar{S}_{ud}$ is the average score of all texts $u = 1, \ldots, U$ and σ_{rd} and σ_{ud} are the sample standard deviation of the reference and unlabeled texts.

So, while Wordfish is more an explorative data analysis tool, Wordscores is closer to a machine learning algorithm. We will not go into further detail of this method but the reader can consult Laver, Benoit, and Garry (2003). A recent method which mixes unsupervised analysis and supervised analysis is called Latent Semantic Scaling (LSS) (Watanabe, 2017) and it uses the vector space models, in the spirit of next section, to reach the goal.

2.4.2 Continuous Space Word Representation: Word2Vec

Many current natural language processing (NLP) systems and techniques treat words as atomic units - there is no notion of similarity between words, as these are represented as indices in a vocabulary. This choice has several good reasons - simplicity, robustness and the observation that simple models trained on huge amounts of data outperform complex systems trained on less data. A very popular model architecture for estimating neural networks language model (NNLM) was proposed in Bengio et al. (2003), where a feedforward neural network[7] with a linear projection layer and a non-linear hidden layer was used to learn jointly the word vector representation and a statistical language model.

In this framework, a statistical model of language can be represented by the conditional probability of the next word given all the previous ones, since

$$\hat{P}(w_1^T) = \prod_{t=1}^{T} \hat{P}(w_t|w_1^{t-1})$$

where w_t is the t-th word, and w_i^j denotes the sub-sequence $w_i^j = (w_i, w_{i+1}, \ldots, w_{j-1}, w_j)$. Such statistical language models have been used in speech recognition, language translation and information retrieval. Thus, n-gram models construct tables of conditional probabilities for the "next word", for each one of a large number of contexts, i.e., combinations of the last $n-1$ words:

$$\hat{P}(w_t|w_1^{t-1}) \simeq \hat{P}(w_t|w_{t-n+1}^{t-1})$$

trying to include variation due to the structure of the language.

The rough idea is that *"The cat is walking in the bedroom"* in the training corpus should help us generalize to make the sentence *"A dog was running in a room"* almost as likely, simply because "dog" and "cat" (resp. "the" and "a", "room" and "bedroom", etc. have similar semantic and grammatical roles. But also
The cat is walking in the bedroom
A dog was running in a room
The cat is running in a room
A dog is walking in a bedroom
The dog was walking in the room
...

The training set is a sequence $w_1, \ldots, w_T$ of words $w_t \in V$, where the vocabulary V is a large but finite set. The objective is to learn the model $f(w_t, \ldots, w_{t-n+1}) = \hat{P}(w_t|w_{t-1})$.

The geometric average of $1/\hat{P}(w_t|w_{t-1})$, also known as *perplexity*, which is the exponential of the average negative log-likelihood. The only constraint on

---

[7]Neural network will be discussed in more detail in Section 2.5.3 but to understand the idea behind this model their precise notion is not essential.

the model is that for any choice of w_{t-1}, $\sum_{i}^{|V|} f(i, w_{t-1}, \ldots, w_{t-n+1}) = 1$, with $f > 0$. By the product of these conditional probabilities, one obtains a model of the joint probability of sequences of words.

The function $f(w_t, \ldots, w_{t-n+1}) = \hat{P}(w_t|w_{t-1})$ is decomposed into two parts:

1. A mapping C from any element i of V to a real vector $C(i) \in R^m$. It represents the distributed feature vectors associated with each word in the vocabulary. In practice, C is represented by a $|V| \times m$ matrix of free parameters.

2. The probability function over words, expressed with C: a function g maps an input sequence of feature vectors for words in context, $(C(w_{t-n+1}), \ldots, C(w_{t-1}))$, to a conditional probability distribution over words in V for the next word w_t. The output of g is a vector whose i-th element estimates the probability $\hat{P}(w_t = i|w_1^{t-1})$; see Figure 2.1

$$f(i, w_{t-1}, \ldots, w_{t-n+1}) = g(i, C(w_{t-1}), \ldots, C(w_{t-n+1})).$$

The function f is a composition of these two mappings (C and g), with C being shared across all the words in the context. Each of these two parts have some parameters. The parameters of the mapping C are simply the feature vectors themselves, represented by a $|V| \times m$ matrix C whose row i is the feature vector $C(i)$ for word i.

The function g may be implemented by a feed forward or recurrent neural network or another parametrized function, with parameters ω. The overall parameter set is $\theta = (C, \omega)$.

Training is achieved by looking for θ that maximizes the training corpus penalized log-likelihood:

$$L = \frac{1}{T} \sum_{i} \log f(w_t, w_{t-1}, \ldots, w_{t-n+1}; \theta) + R(\theta),$$

where $R(\theta)$ is a regularization term. For example, R is a weight decay penalty applied only to the weights of the neural network and to the C matrix. The so-called *softmax* output layer guarantees positive probabilities summing to 1:

$$\hat{P}(w_t|w_{t-1}, \ldots, w_{t-n+1}) = \frac{e^{y_{w_t}}}{\sum_i e^{y_i}}.$$

For full details and examples on NNLM method see Bengio et al. (2003).

The evolution of NNLM is called **Word2Vec** (Mikolov, Sutskever, et al., 2013, Mikolov, Chen, et al., 2013a, Mikolov, Yih, and Zweig, 2013). This method tries to maximize accuracy of vector operations by developing new model architectures that preserve the linear regularities among words taking into account not only previous words but also future words (in the sense of

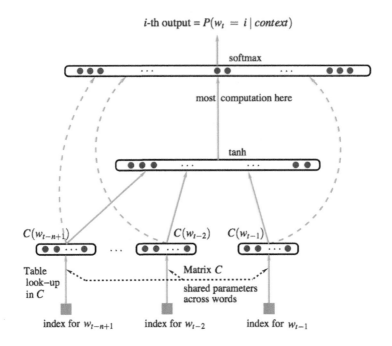

i-th output $= P(w_t = i \,|\, context)$

Figure 2.1
Neural architecture of NNML: $f(i, w_{t-1}, \ldots, w_{t-n+1}) = g(i, C(w_{t-1}), \ldots, C(w_{t-n+1}))$ where g is the neural network and $C(i)$ is the i-th word feature vector. Source: Figure 1 from Bengio et al. (2003).

index t). Somewhat surprisingly, it was found that similarity of word representations goes beyond simple syntactic regularities. Using a word offset technique where simple algebraic operations are performed on the word vectors, it was shown for example that

$$\text{vector(``King'')} - \text{vector(``Man'')} + \text{vector(``Woman'')}$$

results in a vector that is closest to the vector representation of the word "Queen". Two models have been proposed, namely the **Continuous Bag-of-Words Model** (CBOW) and the **Continuous Skip-gram Model**.

The CBOW model is similar to the feedforward NNLM, where the non-linear hidden layer is removed and the projection layer is shared for all words (not just the projection matrix); thus, all words get projected into the same position (their vectors are averaged). It is a **bag-of-words** model as the order of words in the history does not influence the projection. It also use words *"from the future"*, which means that it takes all the other words in the proximity of a given word. The training criterion is to correctly classify the **current (middle) word**. See left panel of Figure 2.2. The Continuous Skip-gram Model is similar to CBOW, but instead of predicting the current word based on the

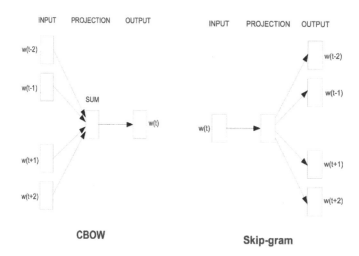

Figure 2.2
Neural Network representation of the CBOW Model (left) and Continuous Skip-gram Model (right). Source: Figure 1 from Mikolov, Chen, et al. (2013b).

context, it tries to maximize classification of a word based on another word in the same sentence. Each **current word** is used as an **input** to a log-linear classifier with continuous projection layer, and it predict words within a certain range before and after the current word. Since the more distant words are usually less related to the current word than those close to it, decreasing weights are used to incorporate this idea. See right panel of Figure 2.2.

2.4.3 Cluster Analysis

Among the unsupervised classification techniques, we can mention text mining methods (Witten, 2004, Feldman and James, 2007) such as the cluster analysis. As by data mining we mean the set of tools aimed at discovering regularities in the data, by text mining we mean the set of techniques able to detect patterns in the texts. Technically speaking, they are the same techniques just applied to any type of data, thus they are methodologically undistinguishable. There are of course some technical issues like the data preparation process, the notion of distance, etc. which make text mining a relevant field of study per se. On final remark is that, while in data mining the information is hidden in the dimensionality of the data, in text mining the information is contained in the texts and it is visibile and transparent though difficult to extract (Hotho, Nurnberger, and Paaß, 2005). To simplify the exposition, we now focus on the general aspects of cluster analysis. This technique is based on the possibility to rearrange observations into homogenous subgroups according to some notion of distance among the data. More precisely, only the notion of dissimilarity is sufficient to perform cluster analysis. A dissimilarity measure d among two objects A and B, i.e., $d(A, B)$, is a function which returns zero when it is calculated on the same object, i.e., $d(A, A) = 0$ for each A (but this does not exclude the possibility of being zero when evaluated for A and B); it is always non-negative, i.e., $d(A, B) \geq 0$, and it is symmetric, that is $d(A, B) = d(B, A)$ for all A and B. If in addition, the triangular inequality holds true we call it a distance (Gordon, 1999). A dissimilarity satisfies the triangular inequality if for any A, B and C, we have $d(A, C) \leq d(A, B) + d(B, C)$. Given a dissimilarity measure $d(\cdot, \cdot)$, clustering algorithms proceed by grouping (agglomerative methods) or splitting (dissociative methods) subsequently the whole set of data. If this procedure is sequential, the method is called hierarchical. For example, an agglomerative hierarchical method is as follows: a first group is formed by taking the closest units in the data. Then each new aggregation occurs either forming a new group of two units, or aggregating a unit to the closest group (according to d) already formed or aggregating two distinct groups. While the distance between two singletons is well defined in general, the distance of a point from a group can be defined in many ways. For example, one can take the average distance of the unit to all the units in a group, or the distance of unit from the frontier of the group, or the minimal distance between this singleton and all the units in the group, etc. Similar problems exists when the task is to define a distance between two groups. A multitude of rules exists in the literature and it is clear that each choice produces a different clustering result. To make the analysis less sensible to the initial choice, and hence more robust in the output, usually researchers perform a meta analysis which consists in running many clustering methods and assessing the stability of the solutions. This approach is called cluster ensembles (Strehl and Ghosh, 2002). Another subtle problem is that the number of clusters is not, in general, the outcome of the cluster analysis but rather a choice of the analyst.

Whichever method of clustering is used, in the end one problem remains, which is: one has to look into the clusters to get some clue of what these clusters mean in term of semantic sense. Supervised clustering methods do exist and one can always force or help the classification by inserting tagged text into the corpus of data and check a posteriori in which cluster these pre-classified data belong to. A primer of clustering in textual analysis with dedicated software is Grimmer and King (2011). Keim (2002) also offers a visual approach to text mining based on related techniques called **Self-Organizing Maps** (SOM) which helps a lot in summarizing similarities and differences among objects lying in different clusters (Kohonen, 2001).

2.4.4 Topic Models

Another approach to classification or grouping which is not strictly cluster analysis is called **Topic Models** (Blei, Ng, and Jordan, 2003, Blei, 2012).

Topic models apply the LDA (Latent Dirichlet Allocation) method and, as the name suggests, it is a way to discover the "topic" of a document. LDA assumes that each text is a mixture of topics. These topics are distributed in the corpus according to some probabilistic distribution (the Dirichlet law). Further, each topic is associated with a sequence of words/stems. The LDA model is in fact a Bayesian model with a two-stage data generation process: first a topic is chosen according to a Dirichlet distribution from the set of possible topics, then a set of words/stems is sampled according to multinomial distribution conditionally on the given topic and the result of this data generating process is an observed text. Statistically speaking, the estimation works in reverse: on the basis of the the words in a document, the models attaches a probability of belonging to some topic. Looking at Figure 2.3 we can see the general data generating process. We have K possible topics and each document may contain $\beta_1, \ldots, \beta_K$ number of topics, where each β_j is distributed according to a Dirichlet distribution (of parameter η) over the words. The n-th word in a document d, denoted by $W_{n,d}$ may belong to more than one topic. The probability of belonging to some topic is given by $Z_{d,n}$ which is the determination of a multinomial distribution that assigns the topic for the n-th word in document d.

Several hyper-parameters exist:

- θ_d is the proportion of topics per document d. This θ_d is also random and distributed according to another Dirichlet distribution of parameter α, and

- η, which is the hyper-parameter for the distribution of the β_j.

As the only observed data are the words, the goal is to estimate the joint distribution:

$$Pr(\text{topics}, \text{proportions}, \text{assignments}|\text{documents}) = p(\beta, \theta, Z|d)$$

and then from this distribution the goal is to assign a topic to each document d, where the joint likelihood is given by

$$p(\beta_{1:K}, \theta_{1:D}, Z_{1:D}, W_{1:D})$$
$$= \prod_{i=1}^{K} p(\beta_j|\eta) \prod_{d=1}^{D} p(\theta_d|\alpha) \left(\prod_{n=1}^{N} p(Z_{d,n}|\theta_d)p(W_{d,n}|\beta_{1:K}, Z_{d,n}) \right)$$

therefore,

$$p(\beta_{1:K}, \theta_{1:D}, Z_{1:D}|W_{1:D}) = \frac{p(\beta_{1:K}, \theta_{1:D}, Z_{1:D}, W_{1:D})}{p(W_{1:D})}.$$

This task is far from being easy, and hence several computational steps are needed, such as: approximate posterior inference algorithms (Metropolis et al., 1953, Hastings, 1970, Robert and Casella, 2005), Mean field variational methods (Blei, Ng, and Jordan, 2003), Expectation propagation (Minka and

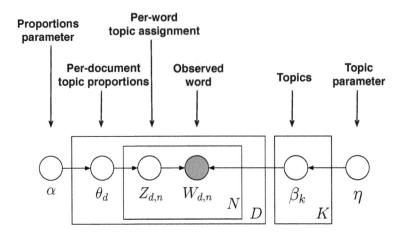

Figure 2.3
The structure of the LDA topics models. Source: Adapted from Figure 1 from Blei, Ng, and Jordan (2003)

Lafferty, 2002), Collapsed Gibbs sampling (Griffiths and Steyvers, 2002), Collapsed variational inference (Teh et al., 2006), Online variational inference (Hoffman, Blei, and Bach, 2010). See also Mukherjee and Blei (2009) and Asuncion et al. (2009).

A variety of different solutions to LDA exist:

- **Correlated Topic Models (CTM)** (Blei and Lafferty, 2007):

    - Draw topic proportions from a logistic normal, where topic occurrences can exhibit correlation;
    - Provide a "map" of topics and how they are related;
    - Better prediction via correlated topics.

- **Dynamic Topic Models** (Blei and lafferty, 2006, Quinn, 2010):

    - LDA assumes that the order of documents does not matter;
    - Not appropriate for corpora that span hundreds of years;
    - We may want to track how language changes over time.

- Wang, Blei, and Heckerman (2008) for a **continuous time variant** of dynamic topic models.

Topic models allow for more general tasks, like authorship identification, etc., and in general they belong to the supervised techniques as the initial set of topics should be explicitated in the process.

There exist also supervised versions of topic models. A few members of this family are:

- **Supervised LDA (SLDA)** (Blei and McAuliffe, 2007), i.e., topics paired with response variables:

  - User reviews paired with a number of stars;
  - Web pages paired with a number of "likes";
  - Documents paired with links to other documents;
  - Images paired with a category.

  Supervised topic models are topic models of documents and **responses**, fit to find topics predictive of the response.

- **Relational topic models (RTM)** (Chang and Blei, 2010), when data sets contain connected observations, e.g.:

  - Citation networks of documents;
  - Hyperlinked networks of web pages;
  - Friend-connected social network profiles.

Finally, it is worth mentioning that very recently Structural Topic Models (STM) were introduced. STM permit to incorporate arbitrary metadata, defined as information about each document, into the topic model. The goal of the Structural Topic Model is to allow researchers to **discover topics** and **estimate their relationship to document metadata**. Outputs of the model can be used to conduct hypothesis testing about these relationships. Like other topic models, the STM is a generative model of word counts. That means we define a data generating process for each document and then use the data to find the most likely values for the parameters within the model. The generative model begins at the top, with document-topic and topic-word distributions generating documents that have metadata associated with them. Within this framework (which is the same as other topic models like LDA), a topic is defined as a mixture over words where each word has a probability of belonging to a topic. And a document is a mixture over topics, meaning that a single document can be composed of multiple topics. As such, the sum of the topic proportions across all topics for a document is one, and the sum word probabilities for a given topic is one. For more details see Roberts et al. (2014).

2.5 Review of Machine Learning Methods

We briefly review some of the most commonly used machine learning methods. This review is clearly far from being complete but gives some insights on why a new perspective should be considered as in Section 2.6.

2.5.1   Decision Trees and Random Forests

Classification and Regression Trees (CART) (Breiman et al., 1984), contrary to, e.g., generalized linear model (GLM), does not assume any parametric structure for the mean $E(Y|X)$ of the response variable Y conditionally on the values of some covariates X, i.e., it is fully *nonparametric*. The final estimation of CART is a ("decision") **tree**. The idea (assumption) behind CART is that the set of covariates X admits a **partition** (a classification into groups) and the tree is simply a representation of this partition. The CART technique is also seen as a **variable selection** method because only variables relevant to the explanation of $E(Y|X)$ remain in the tree. In contrast with many parametric models, **interactions** between variables are handled/discovered automatically. Moreover, data with missing values for some of the covariates can be treated in a large number of cases. The two versions of the model are called: *classification* trees, for categorical Y and *regression* trees for continuous Y.

Let us see an example. Consider a dichotomous variable Y (YES/NO). Generic categorical variables can be used in the same way. When the response variable Y is continuous the only change consists of the loss function used to select the split of the data. As in Clark and Pregibon (1992), we consider one-step look-ahead tree construction with binary splits. Given the set X, we say that its partition can be represented as a tree or, better, by its **terminal nodes** called *leaves*. In the first step, all the observations constitute a unique group. The group is then split into two subgroups by partitioning with respect to one covariate X_j, e.g., into the subset of observations such that $X_j > \tau$ and the subset of observations such that $X_j < \tau$ (for simplicity we assume numerical X_j, but similar ideas apply to categorical or ordinal variables). The variable X_j is chosen among the k covariates $X_1, X_2, \ldots, X_k$ in such a way that the reduction in inhomogeneity inside each node is the maximum achievable. More precisely, this means that, at each step, each variable is considered for a split and all possible splits are considered for such a variable. For each of these binary splits the inhomogeneity of the response variable Y is considered. The split that leads to two subgroups with the minimum inhomogeneity is chosen.

Tree construction is stopped when a minimum number of observations per leaf is reached or when the additional increase in homogeneity is too small. At the end of the process the tree contains nodes and leaves. Each leaf is a subset of X that is as homogeneous as possible with respect to Y. The set of all leaves is the partition of X.

Consider the example in Figure 2.4: this is a classification tree explaining the risk of "having a heart attack" on the basis of age, gender and level of cholesterol based on fictitious data. More formally, assume there are $i = 1, 2, \ldots, C$ classes (values of Y). Let $L(i,j) = 1$ if $i \neq j$; $L(i,i) = 0$, a dissimilarity; A a node; n_i, n_A resp. the number of sample observations belonging to class i and to node A; $\tau(x)$ the true class/response for an observation with covariate vector x; $P(\tau(x) = i) = n_i/n$ (assumption); $\tau(A)$ the class assigned to A (when it is a terminal node); $P(A) = P(x \in A) = \sum_{i=1}^{C} P(x \in A|\tau(x) =$

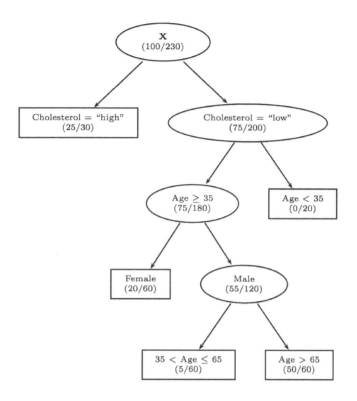

Figure 2.4
$Y = $ heart attack $= 1/0$; $n = 230$ observations and 100 such that $Y = 1$.
Source: Clark and Pregibon (1992).

$i) \cdot n_i/n$: probability of belonging to node A for a new observation with co-variates x; $p(i|A) = P(\tau(x) = i|x \in A)/P(x \in A)$ for future observations; $R(A) = \sum_{i=1}^{C} p(i|A)L(i, \tau(A))$, the risk of node A.

If node A is split into two nodes A_R (right) and A_L (left) then it is always true that:

$$P(A_L)R(A_L) + P(A_R)R(A_R) \leq P(A)R(A).$$

Given that it is always true that if node A is split into two nodes A_R and A_L we have

$$P(A_L)R(A_L) + P(A_R)R(A_R) \leq P(A)R(A)$$

How to choose the best split? Let $I(A)$ the **impurity** of node A

$$I(A) = \sum_{i=1}^{C} f(p_{iA})$$

where p_{iA} is the proportion of observations in A which belong to class i. The function $f(\cdot)$ is called *impurity* function and is such that $f(0) = f(1) = 0$, so that $I(A) = 0$ if all observations belong to a single class, i.e., all $p_{iA} = 0$ but for one i. For example, common impurity functions are the Gini index $f(p) = p(1 - p)$ or the information index $f(p) = -p\log(p)$. The best split to choose is the one with maximal impurity reduction ΔI

$$\Delta I = P(A)I(A) - P(A_L)I(A_L) - P(A_R)I(A_R)$$

The next problem to which extent one should grow a tree? Let $T_1, T_2, \ldots, T_k$ be the terminal nodes of a tree T. Denote by $|T|$ the number of terminal nodes (interpreted similarly to the concept of degrees of freedom in regression)

$$R(T) = \sum_{j=1}^{k} P(T_i)R(T_i)$$

Let $\alpha \in (0, +\infty)$ be the cost of adding a new terminal node to the model, called **complexity**. Let us define

$$\text{cost of tree}: R_\alpha(T) = R(T) + \alpha|T|$$

Let T_α the sub-tree with minimal cost. Then, $T_0 =$ the full tree T, T_∞ is the tree with no splits. A sub-tree is obtained by dropping terminal nodes from the whole tree T. It can be proved that: if T_1 and T_2 are sub-trees of T such that $R_\alpha(T_1) = R_\alpha(T_2)$, then either T_1 is sub-tree of T_2 and therefore $|T_1| < |T_2|$, or vice versa; if $\alpha > \beta$ then either $T_\alpha = T_\beta$ or T_α is a sub-tree of T_β. Therefore, it is possible to define uniquely T_α as the smallest sub-tree of T for which R_α is minimized.

Regression trees work in the same way, but other impurity measures (like deviance) are used for deciding a split. However, CART has limitations:

- the tree tends to overfit the data if they are grown too much, i.e., low bias and high variance, therefore they are not good for prediction. Cross-validation is possible of course

- when there are *too many variables* (for example, in textual analysis) or *too many observations* (social media data), the algorithm becomes unfeasible.

For these reasons the **Random Forests**(RF) algorithm was proposed. Random Forests belong to the *ensemble learning* methods. It essentially consists of building a large number of tree classifier which selects as output (prediction) the **mode** (in case of classification) or the **mean** (in case of regression) of the prediction of each generated tree. This avoids overfitting in the first place.

Random forests generates trees using two ideas: **bagging** (sampling of the variables to use for the trees at each split) and **bootstrap** (sampling on the observations) to reduce computational burden and overfitting. This allows to estimate also the uncertainty of the prediction for regression trees like in bootstrap analysis. Random Forests is also robust to noise, outliers and missing data. It provides also an importance measure for each variable (see original paper, Breiman (2001)).

RandomForests work as follows: let (X, Y) the data with Y the response variable and X the covariates. Let $b = 1, \ldots, B$ the number of bagging replications. Let (X_b, Y) be one bagging replication where X_b is a subset of X. For each b, sample with replacement (bootstrap) n observations and fit a tree T_b. The RF prediction for a new x is the average (or the mode) of the prediction $f_b(x)$ of each tree T_b, i.e.,

$$\hat{f}_B = \frac{1}{B} \sum_{b=1}^{B} f_b(x)$$

This reduces variance without increasing bias. The estimate of the uncertainty of the prediction is evaluated as follows:

$$\sigma = \sqrt{\frac{\sum_{b=1}^{B} (f_b(x) - \hat{f}_B)^2}{B - 1}}$$

In fact, it is more complex: in **each** bagging tree, at **each split**, a subset $X_{b'}$ is selected. This avoids correlation between trees avoiding the most important variables to be selected in all trees. This approach is called "random feature bagging".

The empirical rule is that, if there are p covariates, $b' = \sqrt{p}$ variables should be sampled for classification trees, and $b' = p/3$ for regression trees. These are "rules of thumb" proposed by the authors.

The number of trees and the minimum size of the nodes to allow a split are tuning parameters: not less than 1,000 trees and not less than five observations per node. For a more comprehensive reading see Hastie, Tibshirani, and Friedman (2008). We will see the application of Random Forests in Section 2.6.

2.5.2 Support Vector Machines

Given a set of training data, each marked as belonging to one or the other of two categories $(+1, -1)$, a **Support Vector Machine** (SVM) is a model that assigns new observations to one category or the other as a non-probabilistic binary linear classifier (Cortes and Vapnik, 1995, Ben-Hur et al., 2001). An

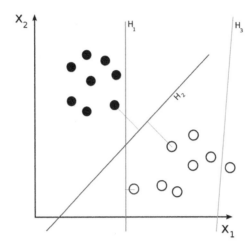

Figure 2.5
The hyperplane H_1 does not separate the classes. H_2 does, but only by a small margin. H_3 separates them by the maximum margin. Source: Wikipedia (2018).

SVM model is a representation of the data as points in a space, mapped so that the observations of the separate categories are divided by a clear gap that is as wide as possible. New data are then mapped into that same space and predicted to belong to a category based on which side of the gap they fall. In addition to performing linear classification, SVMs can efficiently perform a non-linear classification using kernels (Cristianini and Shawe-Taylor, 2000).

Suppose some given data points each belong to one of two classes, and the goal is to decide which class a new data point will be in. In the case of support vector machines, a data point is viewed as a p-dimensional vector, and we want to know whether we can separate such points with a $(p-1)$-dimensional hyperplane. This is called a **linear** classifier.

As there are many hyperplanes that might classify the data one reasonable choice as the best hyperplane is the one that represents the **largest separation**, or margin, between the two classes. So we choose the hyperplane so that the distance from it to the nearest data point on each side is maximized. Figure 2.5 is a graphical representation of separating hyperplanes in two dimensions.

If such a hyperplane exists, it is known as the maximum-margin hyperplane and the linear classifier it defines is known as a maximum margin classifier.

It often happens that the sets to discriminate are not linearly separable in the original space. For this reason, they are mapped into a higher-dimensional space, making the separation easier in that new space. SVMs are designed to ensure that dot products of input data vectors pairs may be computed easily in terms of the variables in the original space, by defining them in terms of a kernel function $k(x, y)$. The hyperplanes in the higher-dimensional space are

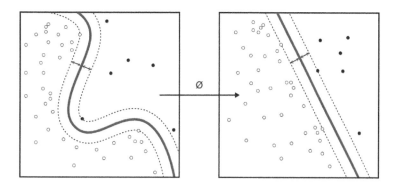

Figure 2.6
Separating hyperplane: kernel approach. Source: Wikipedia (2018).

defined as the set of points whose dot product with a vector in that space is
constant, where such a set of vector is an orthogonal (and thus minimal) set
of vectors that defines a hyperplane.

The vectors defining the hyperplanes can be chosen to be linear combinations with parameters α_i of feature vectors x_i that occur in the original space.
With this choice of a hyperplane, the points x in the **feature space** that are
mapped into the hyperplane are defined by the relation:

$$\sum_i \alpha_i k(x_i, x) = \text{constant}.$$

Kernels are such that $k(x, y)$ becomes small as y goes away from x. Each term
in the above sum measures the degree of closeness of the test point x to the
corresponding observed data point x_i. In this way, the sum of kernels above
can be used to measure the relative nearness of each test point to the data
points originating in one or the other of the sets to be discriminated. Figure 2.6
shows how kernel can help in finding a liner hyperplane in higher-dimensional
space. More formally, given n points of the form (x_i, y_i), $i = 1, \ldots, n$, where x_i
represents the feature p-dimensional vector (covariates) for the observation i
and y_i is either $+1$ or -1, indicating which group the observation i belongs to,
the aim is to find the "maximum-margin hyperplane" that divides the group
of points into two groups.

Any hyperplane can be written as the set of points x satisfying

$$w \cdot x - b = 0,$$

where w is the (not necessarily normalized) normal vector to the hyperplane. The quantity $\frac{b}{\|w\|}$ determines the offset of the hyperplane from the origin along the normal vector w.

If the training data is linearly separable, we can select two parallel hyperplanes that separate the two classes of data, so that the distance between them is as large as possible. The region bounded by these two hyperplanes is called the **margin**, and the maximum-margin hyperplane is the hyperplane that lies halfway between them. With normalized or standardized data, these hyperplanes can be described by the following equations:

$$w \cdot x - b = +1$$

i.e., anything on or above this boundary is of one class, with label 1, and

$$w \cdot x - b = -1$$

i.e., anything on or below this boundary is of the other class, with label -1. The **distance** between these two hyperplanes is $\frac{2}{\|w\|}$, so to maximize the distance between the planes we want to minimize $\|\vec{w}\|$ (see Figure 2.7). Maximum-margin hyperplane and margins for an SVM trained with samples from two classes. Samples on the margin are called the **support vectors**.

To prevent data points from falling into the margin, the following constraint is added to the minimization problem: for each i either

$$w \cdot x_i - b \geq +1, \quad \text{if} \quad y_i = +1$$

or

$$w \cdot x_i - b \leq -1, \quad \text{if} \quad y_i = -1$$

which can be rewritten as

$$y_i(w \cdot x_i - b) \geq 1, \quad \text{for all } 1 \leq i \leq n$$

and then **solve the minimization problem**

$$\text{minimize} \quad \|w\| \quad \text{subject to} \quad y_i(w \cdot x_i - b) \geq 1, \quad \text{for} \quad i = 1, \ldots, n$$

Once w and b are obtained, for a new x the classifier returns $\hat{y} = \text{sgn}(w \cdot x - b)$.

Normally data are not perfectly separable, then the *hinge loss function* is introduced

$$\max\left(0, 1 - y_i(w \cdot x_i - b)\right)$$

which is 0 only for the points that are on the correct side of the margin;

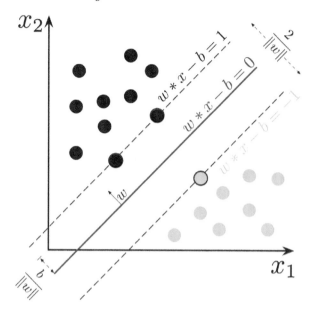

Figure 2.7
Separating hyperplane: kernel approach. Source: Wikipedia (2018).

otherwise it is proportional to the distance from the wrong margin. The minimization problem becomes

$$\left[\frac{1}{n} \sum_{i=1}^{n} \max \left(0, 1 - y_i(w \cdot x_i - b)\right) \right] + \lambda \|w\|^2$$

where the parameter λ determines the tradeoff between increasing the margin-size and ensuring that each point x_i lies on the correct side of the margin. The resulting margins are called "soft margins".

For **non-linear classification**, the **dot products** in the formulas are replaced by some suitable **kernel** function. We will see the application of Support Vector Machines in Section 2.6.

2.5.3 Artificial Neural Networks

An **Artificial Neural Network** (ANN) is a computational model that is inspired by the way biological neural networks in the human brain process information (Anderson, 1995, Haykin, 1999). ANN have generated a lot of excitement in Machine Learning research and industry, thanks to many breakthrough results in speech recognition, computer vision and text processing. Figure 2.8 is a graphical comparison between the biological model and its mathematical counterpart that we call ANN. The basic unit of computation in a neural network is the **neuron**, often called a node or unit. It receives

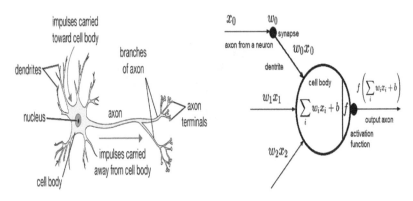

Figure 2.8
A cartoon drawing of a biological neuron (left) and its mathematical model
(right). Source: Karpathy (2018).

input from some other nodes, or from an external source and computes an
output. Each input has an associated **weight** (w), which is assigned on the
basis of its relative importance to other inputs plus some **bias** b (the weight
associated with the bias is 1). The node applies an **activation function** $f(\cdot)$
to the weighted sum of its inputs (see again Figure 2.8). The activation func-
tion $f(|cdot)$ is non-linear and its purpose is to introduce non-linearity into
the output of a neuron. This is important because most real world data is
non-linear and the aim is to train the neurons to learn these non-linear representations. Every activation function (or non-linearity) takes a single number
and performs a certain fixed mathematical operation on it. There are several
activation functions you may encounter in practice:

- *sigmoid*: takes a real-valued input and squashes it to range between 0 and 1

$$f(x) = \sigma(x) = \frac{1}{1 + \exp^{?x}};$$

- *hyperbolic tangent* or *tanh*: takes a real-valued input and squashes it to the
  range $[-1, 1]$

$$f(x) = \tanh(x) = 2\sigma(2x)?1;$$

- *Rectified Linear Unit* or *ReLU*: takes a real-valued input and thresholds it
  at zero (replaces negative values with zero)

$$f(x) = max(0, x);$$

just to mention a few. The main function of bias is to provide every node with
a trainable constant value (in addition to the normal inputs that the node
receives), like in a regression model.

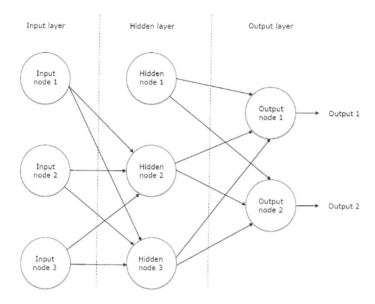

Figure 2.9
A multi-layer Neural Newtwork. Source: Adapted from Figure 3 in Ujjwal
(2016).

For example, we can interpret $\sigma(\sum iw_i x_i + b)$ as the probability of one of the
classes $P(y_i = 1?x_i; w)$ and the probability of the other class would be $P(y_i =
0?x_i; w) = 1?P(y_i = 1?x_i; w)$, since their sum is one. With this interpretation,
the resulting neural network, once optimized for w and b, becomes the so-
called binary Softmax classifier also known as logistic regression. Since the
sigmoid function is restricted to be between 0 and 1, the predictions of this
classifier are based on whether the output of the neuron is greater than 0.5.

Figure 2.9 represents a feedforward multi-layer neural network. A feedforward neural network can consist of three types of nodes:

- **input nodes:** the input nodes provide information from the outside world to
  the network and are together referred to as the *input layer*. No computation
  is performed in any of the input nodes, they just pass on the information to
  the hidden nodes;

- **hidden nodes**: the hidden nodes have no direct connection with the outside
  world (hence the name "hidden"). They perform computations and transfer
  information from the input nodes to the output nodes. A collection of hidden
  nodes forms a *hidden layer*. While a feedforward network will only have a
  single input layer and a single output layer, it can have zero or multiple
  hidden layers;

- **output nodes**: the output nodes are collectively referred to as the *output layer* and are responsible for computations and transferring information from the network to the outside world.

In a feedforward network, the information moves in only one direction, forward, from the input nodes, through the hidden nodes (if any) and to the output nodes. There are no cycles or loops in the network, like in, for example, Recurrent Neural Networks, in which the connections between the nodes form a cycle. The ANN in Figure 2.8 is called **single layer perceptron** because it does not contain any hidden layer. The ANN depicted in Figure 2.9 is called **multi layer perceptron** (MLP) because it has one or more hidden layers. MLP are the most used types of ANN today thanks to the availability of computer power.

Then ANNs have a plethora of different names according to their structure or combination and to the way optimization is done. For example, backpropagation is one way of setting up an optimization algorithm. We do not go into detail here. These models allow to capture highly non-linear relationships between the input and the output but are also high-dimensional in terms of number of statistical parameters and tuning parameters (number of layers, number of neurons per layer, activation function, etc.).

Convolutional Neural Networks (CNN) (Fukushima, 1979, 1980, 2013) and the wider class of Deep Neural Networks (Hochreiter, Younger, and Conwell, 2001) are now trendy and modern versions of artificial neural networks that are usually called Deep Learning methods. They are in fact multi-layered, fully connected artificial networks which became popular due to the increase in computing capabilities of recent years. They prove to be very accurate in classifying images and in vision application in general. We do not discuss them in detail here. A full discussion can be found in Schmidhuber (2014).

2.6   Estimation of Aggregated Distribution

As seen, the "traditional" machine learning methods (e.g., multinomial regression, Random Forests (RF), Support Vector Machines (SVM), Artificial Neural Networks (ANN), etc.) use the individual hand coding from the training set to construct a model $P(D|S)$ for $P(D)$ as a function of S to train a model that predicts the outcome of $\hat{d}_j = D$ for the texts with $S = s_j$ belonging to the test set. Then, when all data have been imputed in this way, these estimated categories $\hat{d}_j$ are aggregated to obtain a final estimate of $\hat{P}(D)$. In matrix form, we can write

$$\underset{M \times 1}{P(D)} = \underset{M \times \bar{K}}{P(D|S)} \underset{\bar{K} \times 1}{P(S)} \tag{2.1}$$

where $P(D)$ is a $M \times 1$ vector, $P(D|S)$ is a $M \times \bar{K}$ matrix of conditional probabilities and $P(S)$ is a $\bar{K} \times 1$ vector which represents the distribution of S_i over the corpus of texts. Once $P(D|S)$ is estimated from the training set with, say, $\hat{P}(D|S)$, then for each document in the test set with stem vector s_j, the opinion $\hat{d}_j$ is estimated with the simple Bayes estimator as the maximizer of the conditional probability, i.e., $\hat{d}_j = \arg\max_{D \in \mathcal{D}} \hat{P}(D|S = s_j)$.

In the naïve nonparametric model, the elements of the matrix $P(D|S)$, e.g., $P(D = D_i|S = S_k)$ are estimated taking the proportion of all texts in the training set hand-coded as $D = D_i$, which have $s_j = S_k$ as stem vector. Any other model (SVM, etc.) will do essentially the same thing in more sophisticated or smarter ways, but the discussion here does not change.

Now, let us take $i \neq 0$; then for most S_j these conditional probabilities are zero as the opinion D_i is expressed only for a small subset of $S_j \in \bar{S}$. On the contrary, if we assume that D_0 is the category under which we confine the "*Off-Topic*" texts, i.e., text in the corpus who do not really matter the analysis and which constitute the noise, then $P(D = D_i|S = S_j) > 0$ for almost all $S_j \in \bar{S}$. This means that, if n is relatively small, the estimation of $P(D|S)$ will be very poor and, most of the times, the predicted category $D = d$ for a text in the test set will be imputed as $d = D_0$, being D_0 the most frequent case in the corpus of texts. As a result, $D = D_0$ will be over estimated and when the aggregation occurs this strong bias will persist so that $P(D)$ will be strongly biased as well.

2.6.1 The Need of Aggregated Estimation: Reversing the Point of View

Following Hopkins and King (2010), the idea is to change point of view and focus on what can be really and accurately estimated. Instead of equation (2.1) one can proceed considering this new equation

$$\underset{\bar{K} \times 1}{P(S)} = \underset{\bar{K} \times M}{P(S|D)} \underset{M \times 1}{P(D)} \tag{2.2}$$

where now $P(S|D)$ is a $\bar{K} \times M$ matrix of conditional probabilities whose elements $P(S = S_k|D = D_i)$ represent the frequency of a particular stem S_k given the set of texts which actually express the opinion $D = D_i$. In this case, all these probabilities can be considerably well estimated if there is a sufficient number of texts[8] in the training set which are hand coded as $D = D_i$. Indeed, if $i \neq 0$, only few S_k's will be observed among those text expressing the opinion D_i, for all the remaining stems in $\bar{S}$ these probabilities will be zero. On the other side, if $i = 0$, all these probabilities $P(S = S_j|D = D_0)$ are almost all close to zero as most of the texts belong to D_0. Then, the solution of the

---

[8]When this is not the case, one should increase the sample size sequentially as explained in Section 2.8.

problem is as follows

$$(\text{inverse problem}) \quad \underset{M \times 1}{P(D)} = [\underset{M \times M}{P(S|D)^T P(S|D)}]^{-1} \underset{M \times \bar{K}}{P(S|D)^T} \underset{\bar{K} \times 1}{P(S)} \quad (2.3)$$

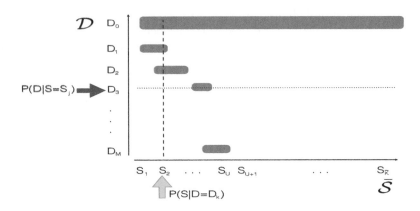

Figure 2.10
The space $\bar{S} \times \mathcal{D}$. The reason why, when the noise D_0 category is dominant in the data, the estimation of $P(S|D)$ is reasonably more accurate than the estimation of counterpart $P(D|S)$.

Figure 2.10 exemplifies the idea of aggregated estimation behind formula (2.3). The plot represents the product space $\bar{S} \times \mathcal{D}$. For simplicity, we have reordered the vectors of stems S_j on the horizontal axis in the way that, say, S_1 to S_U which matter the categories D_1 to D_M appear first, and from S_{U+1} to $S_{\bar{K}}$ the remaining vector of stems. Clearly, the vectors S_1 to S_U sometimes also belong to D_0 and do not necessarily appear uniquely for each D_k. From this simple representation, one can see that, for example, D_2 is supported by a limited number of stems S, while D_0 essentially by all stems. This makes $P(D = D_0|S_j) >> P(D = D_k|S_j)$ for all $k \neq 0$ and any given S_j (horizontal dotted line). On the contrary, say, $P(S = S_2|D = D_k)$ will be non-zero only for D_0, D_1 and D_2 (vertical dashed line in the example of Figure 2.10.)

Notice that equation (2.2) is very close to the regression analysis model $Y = X\beta$ and (2.3) to its classical least squares solution $\hat{\beta} = [X^T X]^{-1} X^T Y$ with $\beta = P(D)$, $X = P(S|D)$ and $Y = P(S)$. The only difference is that the solution (2.3) via linear regression analysis does not necessarily produce non-negative estimates $\beta_j \geq 0$, $j = 0, 1, \ldots, M$, such that $\sum_{j=0}^{M} \beta_j = 1$, therefore it is possible to use simple quadratic programming (QP) to solve the task.

Remark that the direct solution of (2.3) does not allow for the individual classification of the d_j's in the training set.

2.6.2 The ReadMe Solution to the Inverse Problem

The other compelling issue is that $P(S|D)$ involves the estimation over L stems, which makes a direct solution of the problem very hard, if not impossible, to solve. For this reason, Hopkins and King (2010), in the ReadMe algorithm, proposed an approach based on the random sampling of the length of the stem vector, i.e., L of a given size and estimate $P(S|D)$ only for these reduced stem length vectors. For each simulation an estimate of $P(D)$ is obtained and the results are averaged over many simulations like in a statistical bagging approach. The result is a slow algorithm where the estimates have usually large variance. ReadMe essentially implements bagging over the explanatory variables (the stems.) Nevertheless, the ReadMe approach is, so far, the only available and pioneer method for aggregated sentiment analysis.

2.7 The iSA Algorithm

By exploiting the ideas in Hopkins and King (2010), iSA is a fast and more accurate implementation of (2.2) which does not require resampling method and uses the complete length of stems by a simple trick of dimensionality reduction using an idea inherited from Coarsened Exact Matching algorithm (Iacus, King, and Porro, 2011). More precisely, each vector of stems, e.g., $s_j = (0, 1, 1, 0, \ldots, 0, 1)$ is transformed into a string $C_j =$ "0110$\cdots$01" and this sequence of 0's and 1's is further translated into hexadecimal notation such that the sequence $C_j =$ '11110010' is recoded as $\lambda_j =$ 'F2' or $C_j =$ '11100101101' as $\lambda =$ 'F2D', and so forth. So each text is actually represented by a single label λ of relatively short length. This dimensionality reduction trick has at least three advantages over the ReadMe approach.

Firstly, the problem has been transformed from one in which the matrix Σ has dimension $N \times \bar{K}$ into a vector of length $N \times 1$ resulting in memory-efficient storage. Moreover, the label ℓ_j representing the sequence s_j of, say, a hundred of 0's and 1's can be stored in just 25 characters, i.e., the length is reduced to one-fourth of the original one.

Secondly, the estimation of $P(S|D)$ is as easy as an instant tabulation over the n texts (or labels ℓ) of the training set which has complexity of order n (and N for $P(S)$).

Finally, the original problem (2.3) is solved directly and exactly (i.e., without simulation) using quadratic programming (QP) to solve $\min(-\mu^T b + \frac{1}{2} b^T Q b)$ where $Q = P(S|D)^T P(S|D)$, $\mu = P(S)^T P(S|D)$, $b = P(D)$ under the constraints that the coefficients are in $[0, 1]$ and sum up to the unit.

2.7.1 Main Advantages of iSA over the ReadMe Approach

it is worth mentioning that the QP approach was proposed in the case of ReadMe by Hopkins and King (2010) as well, to solve each version of the problem (2.3) during the bagging step. QP is not the essential part of the procedure as it can be replaced by any other algorithm to solve (2.3), i.e., constrained linear regression.

In the case of iSA, the reduction of dimensionality allows for instantaneous estimation of $P(S|D)$ and therefore a single QP step is enough to obtain the solution.

Further, it is not clear whether the resampling method proposed for ReadMe throwing away information at each step, really guarantees convergence of the estimates when the dimensionality of the set S grows too much. ReadMe is indeed theoretically unbiased in each replication, but the variability of the estimates is quite large due to bagging. A further problem of ReadMe, also related to bagging, is that if the number of categories D is larger than 10–12, the convergence of the algorithm is much slower or the algorithm (at least the official release available in the R package ReadMe, Hopkins and King (2013)) does not converge even after tweaking on some parameters.

In the case of iSA the solution of the problem (2.3) is direct, therefore plain bootstrap can be used to obtain correct standard errors, resampling on the observations (the texts) and not on the explanatory variables (the stems). Due to the fact that bagging is not used, iSA converges even when the dimensionality of $\mathcal{D}$ is very large. Indeed, iSA is generally stable.

To summarize, iSA is an extremely fast, memory-efficient, stable and accurate algorithm for which the standard errors of the estimates are validated by standard bootstrap theory without any particular assumption or additional proof.

2.8 The iSAX Algorithm for Sequential Sampling

The dimensionality reduction used by iSA has the effect of assuming a reduced variability in the configuration of stems, which means that if a vector of stems in the test set is not observed in the training set, the corresponding entry of $P(S|D)$ will be estimated as zero. This is not a problem when the training set is sufficiently large, i.e., n is large, and randomly selected from the entire corpus of texts thanks to the law of large numbers.

But in operational setups the tagging of texts occurs in a sequential way, i.e., the coders usually skip many texts in the finding of those texts which contain the relevant categories D or a filter is applied by the researcher to a random set of texts before sending them out for manual tagging. This results in a sequential procedure that gives a training set which is not obtained by simple random sampling. In this situation, most machine learning methods of individual classification will fail in the presence of training sets of small sample size, due to the mismatch of representativity of the training set with respect to the test set. This is the second reason why the ReadMe approach samples randomly the set of stems in the attempt to get a better estimate of $P(D)$. Indeed, to avoid bias in the estimates and increase accuracy, a sufficient number of hand-coded texts is needed per each opinion $D \in \mathcal{D}$. Empirical evidence seems to require at least 20 tagged texts per category D.

On the contrary, it is still possible to obtain the same goal in the presence of sequential sampling by modifying the iSA algorithm as follows: The sequence λ of hexadecimal codes is split into subsequences of length 5, which corresponds to 20 stems in the original 0/1 representation. For example, consider the sequence λ_j = 'F2A10DEFF1AB4521A2' of 18 hexadecimal symbols and the tagged category $d_j = D_3$. The sequence λ_j is split into $4 = \lceil 18/5 \rceil$ chunks of length five or less: λ_j^1 = 'aF2A10', λ_j^2 = 'bDEFF1', λ_j^3 = 'cAB451' and λ_j^4 = 'eA2'. We denote by λ' these split λ's in Figure 2.11. At the same time, the d_j are replicated (in this example) four times, i.e., $d_j^1 = D_3$, $d_j^2 = D_3$, $d_j^3 = D_3$ and $d_j^4 = D_3$. The same applies to all sequences of the training set and those in the test set. This method results into a new data set of length which is four times the original length of the data set, i.e., $4N$. To this new data set, the standard iSA can be applied and the problem (2.3) is solved in one iteration of QP with the same complexity as before. Clearly, in this approach there is no need to average the estimates like in ReadMe as there will be one single solution only. Again, bootstrap on the vector of stems can be used to obtain standard errors of the new iSA estimates.

Notice that this new version of iSA works almost equally well in the case of truly random training set, but also works in the general applied framework of sequential coding. This version of iSA has larger variability than the original version presented in Section 2.7. To distinguish between the two versions of iSA, we denote the latter as iSAX. Figure 2.11 summarizes the different steps of the iSA/iSAX algorithm.

2.9   Empirical Comparison of Machine Learning Methods

To describe the performance of iSA, we compare this new algorithm with ReadMe, the direct competitor of aggregated sentiment analysis available in the R package `ReadMe` (Hopkins and King, 2013). We also consider two other classical supervised machine learning methods: the (RF) Random Forest method (Breiman, 2001) available in the R package `randomForest` (Liaw and Wiener, 2002), and Support Vector Machines (SVM) with spherical kernel as implemented in the R package `e1071` (Meyer et al., 2014). We do not

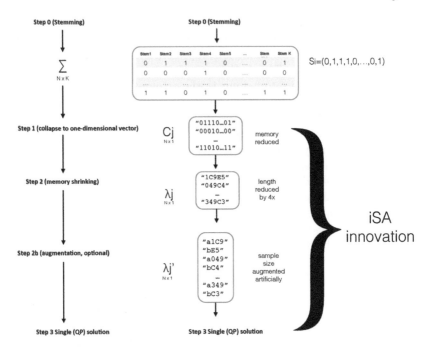

Figure 2.11
The iSA/iSAX algorithm. Source: Ceron, Curini, and Iacus, (2016).

include artificial neural newtworks in the competition as there are a plethora of them that can be chosen (each with its own tuning parameters), although in Figure 2.12 we show some comparison of the above methods also with Convolutional Neural Networks (CNNs) (Fukushima, 1979, 1980, 2013) and Deep Neural Networks (Hochreiter, Younger, and Conwell, 2001). The figure shows the relative efficiency of iSA/iSAX with respect to the other methods in terms of training set sample size and estimation bias.

For each of the data sets presented in this section, we run a simulation experiment taking into account only the original training set of n observations. The experiment is designed as follows: we randomly partition the n observations into two portions: $p \cdot n$ observations will constitute a new training set and $(1 - p) \cdot n$ observations are considered as test set, i.e., the true category is disregarded. We let p vary in 0.25, 0.5, 0.75 and 0.9.

In this way it is possible to evaluate the performance of each classifier. Indeed, we estimate $\hat{P}(D)$ for all texts (in the training and test sets) using iSA, RF, SVM and ReadMe and we compare the estimated distribution using

MAE (mean absolute error), i.e.,

$$MAE(\text{method}) = \frac{1}{M} \sum_{i=1}^{M} \left| \hat{P}_{\text{method}}(D_i) - P(D_i) \right| \tag{2.4}$$

and the χ^2 (Chi-Squared) test

$$\chi^2(\text{method}) = \sum_{i=1}^{M} \frac{(\hat{P}_{\text{method}}(D_i) - P(D_i))^2}{P(D_i)}, \tag{2.5}$$

where the "method" is one among iSA, iSAX, RF, SVM and ReadMe. The tables report also the estimation done via iSAX, a modification of iSA designed for sequential sampling, described in Section 2.8. We run each experiment 100 times.[9] All computations have been performed on a MacBook Pro, 2.7 GHz with Intel Core i7 processor and 16 GB of RAM. All times for iSA include 100 bootstrapping replications for the standard error of the estimates even if these estimates are not shown in the Monte Carlo analysis. For the analysis we use Martin Porter's stemming algorithm and the `libstemmer` library from `http://snowball.tartarus.org` as implemented in the R package `SnowballC` (Bouchet-Valat, 2014). After stemming, we drop the stems whose sparsity index is greater than the q% threshold, i.e., stem, which appear less frequently than q% in the whole corpus of texts. Stop words, punctuation and white spaces are stripped as well from the texts.

We show the analysis [10] of the so called "Large Movie Review Dataset" (Maas et al., 2011), originally designed for a different task. This dataset consists of 50,000 reviews from IMDb, the *Internet Movie Database* (`http://www.imdb.com`) manually tagged as positive and negative reviews but also including the number of "stars" assigned by the internet users to each review. Half of these reviews are negative and half are positive. Our target set of categories $\mathcal{D}$ consists of the number of stars assigned to each review, a much difficult task than the dicothomus classification into positive and negative. The true target distribution of stars $P(D)$ is given in Table 2.2. Categories "5" and "6" do not exist in the original data base. We have $M = 8$ for this dataset. The original data can be downloaded at `http://ai.stanford.edu/~amaas/data/sentiment/`. For the simulation experiment we confine the attention to the 25,000 observations in the original training set. Notice that in this data set there is no miss-specification or *Off-Topic* category, so we should expect that traditional machine learning method to perform well.

As can be seen from Table 2.2, the reviews are polarized and the true distribution of $P(D)$ is unbalanced: D_1 and D_{10} amount to the 40% of the total

---

[9]A larger number of simulations is unfeasible in most cases given the unrealistic computational times of the methods other than iSA.

[10]The results of this analysis are taken from Ceron, Curini, and Iacus, (2016). We refer the reader to this paper for further examples.

Table 2.2
(Top) True distribution of $P(D)$ for the Large Movie Review dataset. Fully hand-coded training set sample size $n = 2,5000$. (Bottom) The distribution $P(D)$ of the random sample of $n = 2,500$ texts used in the simulation studies of Table 2.3. Source: Ceron, Curini, and Iacus, (2016).

N. stars (D)	1	2	3	4	7	8	9	10	Total
Target $P(D)$	20.4%	9.1%	9.7%	10.8%	10.7%	12.0%	9.1 %	18.9%	100%
Tagged texts	5,100	2,284	2,420	2,696	2,496	3,009	2,263	4,732	$n = 25,000$
Target $P(D)$	18.9%	9.9%	9.3%	11.2%	9.8%	12.5%	8.9 %	19.5%	100%
Tagged texts	355	186	174	210	184	234	166	366	$n = 2,500$

probability mass distribution, the remaining being essentially equi-distributed. Still, this case is not within the assumption of iSA or ReadMe and one should expect a good performance of SVM and RF given that there is no miss classification and we are performing random sampling to select the training set. The reason why this happens is related to the over-representation of extreme categories (like D_1 and D_{10}) which take the role of the D_0 category.

After elementary stemming and removing stems with sparsity index of 0.95, the remaining stems are $L = 320$, still a huge number of predictors for both SVM and RF which make the computational time for them very high. To reduce the computational times, we considered a random sample of size 2,500 observations from the original training set of 25,000.

The results of the analysis are collected in Table 2.3. In this example, both ReadMe and iSA outperform the other methods from small to medium training set sample sizes ($p = 25\%, 50\%$). The algorithms iSA and iSAX outperform all methods in terms of MAE and χ^2. All methods, but ReadMe[11], behave as expected as the sample size increases, i.e., the MAE and χ^2 decrease, as well as the Monte Carlo standard deviation of the MAE estimate (in parentheses.)

Notice also that while the computational times remain essentially stable and around fraction of seconds for iSA and half minute for ReadMe, for the other two methods the computational times grow more than linearly with the number of observations. For example, with RF, if we pass from $p = 0.25$ to $p = 0.50$, i.e., we double the size of the training set, the time increases by an amount of $2.17\times = 35.7/16.4$; if we pass from $p = 0.25$ to $p = 0.75$, i.e., we triple the size of the training set, the computational time increases by a factor of $3.45\times = 56.7/16.4$. To summarize, for all p's the iSA algorithm is faster, more stable and more accurate than all other competitors. iSAX is also very accurate with a slightly larger variability and still much better than ReadMe,

---

[11]This might be due to the fact that, increasing the sample size of the training set the number of stems on which ReadMe has to perform bagging increases as well. In some cases, the algorithm does not provide stable results as the number of re-sampled stems is not sufficient and therefore, an increased number of bagging replications will be necessary. In our simulations we kept all tuning parameters fixed and we changed the sample size only.

Table 2.3
Monte Carlo results for the Large Movie Review dataset. Table contains MAE, Monte Carlo standard errors of MAE estimates, χ^2 statistic and execution times for each individual replication in seconds as multiple of the base line which is iSAX. Sample size $N = 2,500$ observations from the original Large Movie Review training set. Number of stems 320, threshold 95%. For the iSAX method we report, in parentheses, the number of seconds for each single iteration in the analysis, which means, the total time of the simulation must be multiplied by a factor of 100. For example, while a complete analysis with iSAX for $p = 25\%$ requires $0.3\,\text{s} \times 100 = 300\,\text{s} = 6\,\text{m}$, for the RF algorithm it requires $6\,\text{m} \times 16.4 = 98.4\,\text{m} = 1\,\text{h}\,39\,\text{m}$ and about $7\,\text{h}$ for $p = 90\%$. Source: Ceron, Curini, and Iacus, (2016).

Method	RF	SVM	ReadMe	iSA	iSAX
$p = 25\%$ ($n = 625$)					
MAE	0.088	0.152	0.040	0.010	0.014
MC Std. Dev.	[0.009]	[0.002]	[0.005]	[0.003]	[0.004]
χ^2	0.287	0.737	0.087	0.005	0.009
speed	(16.4×)	(3.0×)	(15.6×)	(0.2×)	(1 = 0.3 s)
$p = 50\%$ ($n = 1,250$)					
MAE	0.060	0.101	0.039	0.006	0.009
MC Std. Dev.	[0.005]	[0.001]	[0.004]	[0.002]	[0.003]
χ^2	0.126	0.344	0.085	0.002	0.004
speed	(35.7×)	(7.2×)	(14.7×)	(0.2×)	(1 = 0.3 s)
$p = 75\%$ ($n = 1,875$)					
MAE	0.030	0.051	0.039	0.003	0.006
MC Std. Dev.	[0.002]	[0.001]	[0.004]	[0.001]	[0.002]
χ^2	0.032	0.100	0.080	0.001	0.002
speed	(56.7×)	(13.7×)	(14.3×)	(0.2×)	(1 = 0.3 s)
$p = 90\%$ ($n = 2,250$)					
MAE	0.012	0.020	0.039	0.002	0.004
MC Std. Dev.	[0.001]	[0.001]	[0.007]	[0.001]	[0.001]
χ^2	0.005	0.019	0.081	0.000	0.001
speed	(69.6×)	(18.2×)	(14.1×)	(0.2×)	(1 = 0.3 s)

RF and SVM. At least in this example, but the results given in Ceron, Curini, and Iacus, (2016) show that they are quite replicable in different datasets.

Given that this dataset is completely hand coded we can use all the 25,000 observations in the original training set and the 25,000 observations of the test set, we can run the four classifier, and compare with the true distribution with the corresponding estimates. For this we disregard the hand coding of the 25,000 observations in the test set. The results given in Table 2.4 show that iSA and iSAX are the most accurate methods in terms of MAE and χ^2, then follow ReadMe, RF and SVM. Nevertheless, for each iteration iSAX took only 5.7 seconds with bootstrap (only 2.6 seconds for iSA) while SVM (resp. RF) took 4,640 (resp. 798.3) seconds, which is more than 800 (resp. 140) times slower than iSAX.

We will show the results on the accuracy of the estimates in Section 2.9.1 where we will consider confidence intervals with or without sequential sampling.

Table 2.4
Classification results on the complete Large Movie Review dataset. The table
contains the estimated distribution of $P(D)$ for each method, the relative
MAE and the computational times in seconds, relative to the classification of
the set of 50,000 observations from the Large Movie Review dataset where
25,000 observations are used as training set. Number of stems 309, threshold
95%. Source: Ceron, Curini, and Iacus, (2016).

$n = 25,000$	RF	SVM	ReadMe	iSA	iSAX
MAE	0.059	0.099	0.044	0.002	0.014
χ^2	0.116	0.329	0.120	0.000	0.010
Time	798.3 s	4,640.9 s	105 s	2.6 s	5.7 s

Table 2.5
Classification results on the complete Large Movie Review dataset. Data as
in Table 2.7 for the whole dataset of 50,000 observations with $n = 25,000$.
Top: the final estimated distributions, bottom: the 95% confidence interval
upper-bound and lower-bound estimates for iSA and iSAX. Source: Ceron,
Curini, and Iacus, (2016).

Stars	True	iSAX	ReadMe	RF	SVM	iSA
1	0.202	0.193	0.194	0.313	0.600	0.204
2	0.092	0.103	0.241	0.046	0.046	0.091
3	0.099	0.108	0.088	0.052	0.049	0.097
4	0.107	0.108	0.103	0.072	0.054	0.108
7	0.096	0.066	0.105	0.062	0.050	0.100
8	0.117	0.115	0.067	0.090	0.060	0.121
9	0.092	0.081	0.110	0.046	0.045	0.090
10	0.195	0.226	0.092	0.318	0.095	0.189
MAE		0.013	0.044	0.059	0.099	0.002
χ^2		0.010	0.120	0.116	0.329	0.000

Stars	Lower	True	iSA	Upper	Stars	Lower	True	iSAX	Upper
1	0.201	0.202	0.204	0.206	1	0.172	0.202	0.193	0.215
2	0.090	0.092	0.091	0.093	2	0.083	0.092	0.103	0.123
3	0.095	0.099	0.097	0.099	3	0.092	0.099	0.108	0.124
4	0.105	0.107	0.108	0.110	4	0.089	0.107	0.108	0.127
7	0.098	0.096	0.100	0.102	7	0.030	0.096	0.066	0.101
8	0.119	0.117	0.121	0.123	8	0.097	0.117	0.115	0.133
9	0.089	0.092	0.090	0.092	9	0.061	0.092	0.081	0.102
10	0.187	0.195	0.189	0.192	10	0.210	0.195	0.226	0.243

A side comment is related to the efficiency in the task of hand coding.
Given that each analysis requires a given training set, it is important to evaluate the impact of this step for large-scale projects like the one considered
in this book. Figure 2.12 shows the reduction of bias (evaluated as MAE)

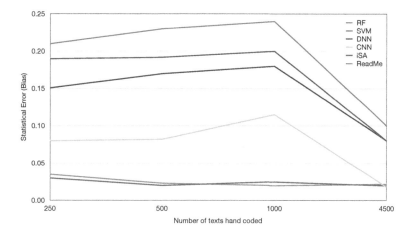

Figure 2.12
The plot shows that to obtain the same level of bias (measured as MAE, vertical axis) iSA requires very small training set (hand-coded texts, horizontal axis) than fancy machine learning algorithms like Deep Neural Networks (DNN), or other classical Convolutional Neural Networks (CNN), or the traditional Support Vector Machines (SVM) and Random Forests (RF). For example, CNN reaches the same level of bias as iSA at 4,500 tagged texts that is 20 times more than iSA which requires 250. Source: Ceron, Curini, and Iacus, (2016).

in the estimation of $P(D)$ for the Large Movie Review dataset as a function of the training set size. In this example the comparison has been extended to include Convolutional Neural Networks (CNNs) (Fukushima, 1979, 1980, 2013) and Deep Neural Networks (Hochreiter, Younger, and Conwell, 2001) which are now trendy versions of artificial neural networks. The plot shows that CNN can perform as good as iSA in terms of bias, provided that a large training set is hand coded. In the Monte Carlo experiment it turned out that 4,500 hand-coded text for CNN provide the same accuracy as iSA (or ReadMe) with 250 hand coded texts. This training set size can be prohibitive in most applications but of course it is not a problem in specific contexts like image classification as done by big players like Google, Bing or Yahoo!

2.9.1 Confidence Intervals

We finally evaluate 95% confidence intervals for iSA and iSAX in both cases in Tables 2.5 and 2.6. The other methods require further bootstrap analysis in order to produce standard errors which make the experiment unfeasible so we didn't consider standard errors for ReadMe, SVM and RF. Tables 2.5 and 2.6 we can see that in all cases but one, both iSA and iSAX confidence intervals contain the true values of the parameters. The only case in which true value is

Stars	True	iSAX	ReadMe	RF	SVM	iSA
1	0.202	0.114	0.128	0.099	0.054	0.126
2	0.092	0.107	0.146	0.085	0.295	0.126
3	0.099	0.116	0.116	0.087	0.097	0.125
4	0.107	0.132	0.107	0.143	0.074	0.124
7	0.096	0.112	0.119	0.069	0.072	0.123
8	0.117	0.130	0.129	0.086	0.086	0.126
9	0.092	0.124	0.135	0.142	0.085	0.124
10	0.195	0.166	0.120	0.288	0.237	0.125
MAE		0.029	0.037	0.045	0.061	0.036
χ^2		0.038	0.059	0.079	0.212	0.051

Stars	Lower	True	iSA	Upper	Stars	Lower	True	iSAX	Upper
1	0.039	0.202	0.126	0.213	1	0.089	0.202	0.114	0.139
2	0.026	0.092	0.126	0.225	2	0.084	0.092	0.107	0.131
3	0.041	0.099	0.125	0.210	3	0.090	0.099	0.116	0.141
4	0.041	0.107	0.124	0.208	4	0.104	0.107	0.132	0.160
7	0.036	0.096	0.123	0.210	7	0.086	0.096	0.112	0.137
8	0.036	0.117	0.126	0.216	8	0.103	0.117	0.130	0.156
9	0.038	0.092	0.124	0.210	9	0.089	0.092	0.124	0.159
10	0.033	0.195	0.125	0.217	10	0.127	0.195	0.166	0.206

Table 2.6
Classification results on the complete Large Movie Review dataset. Data as in Table 2.7 for the whole dataset of 50,000 observations with $n = 80$. Top: the final estimated distributions, bottom: the 95% confidence interval upper-bound and lower-bound estimates for iSA and iSAX.

$n = 25,000$	RF	SVM	ReadMe	iSA	iSAX
MAE	0.059	0.099	0.044	0.002	0.014
χ^2	0.116	0.329	0.120	0.000	0.010
Time	798.3 s	4640.9 s	105 s	17.2 s	41.8 s
$n = 80$	RF	SVM	ReadMe	iSA	iSAX
MAE	0.045	0.061	0.037	0.036	0.029
χ^2	0.079	0.212	0.059	0.050	0.038
Time	14.5 s	2.5 s	114.5 s	15.6 s	40.5 s

Table 2.7
Classification results on the complete Large Movie Review dataset. The table contains the estimated distribution of $P(D)$ for each method, the relative MAE and the computational times in seconds, relative to the classification of the set of 50,000 observations from the Large Movie Review dataset where 25,000 observations are used as training set (top) and (bottom) where only 10 observations per category have been chosen for the training set (sample size: training set = 80, test set = 49,840). A total of 1,000 bootstrap replications for the evaluation of the standard errors of iSA and iSAX estimates. Number of stems 309, threshold 95%. Source: Ceron, Curini, and Iacus, (2016).

outside the confidence interval is the category D_1 = "1 Star" in Table 2.6 for iSAX and $n = 80$, which is, as seen in all other tables, an extremely difficult case for all methods.

2.10 Conclusions

This chapter has reviewed the most commonly used techniques of text analysis and in particular machine learning methods. The main statistical evidence from both the theoretical and empirical results is that for the application to social media data, or web data in general, given the scale of the data and the noise in the data, it is better to focus on the estimation of the aggregated distribution of opinions rather than on the individual classification of each text. The only two competitors available are ReadMe and iSA, the latter being more efficient in term of bias, variance, number of categories (dimension of D) and computer time.

Of course machine learning methods for individual classification are perfectly fine for the scope they have been designed and when the classical assumptions of statistical inference hold: random sampling, correctly specified reference population, no noise in the data, large training set. Unfortunately big data coming from social media are not as such in that the reference population is often miss-specified, data contain noise and tagging is made sequentially, i.e., tagging is continued until a certain number of data are correctly hand coded for each opinion and the length of this process is a function of the level of noise in the data: the higher the noise, the more the need to hand code. So for this project of well-being measurement we will adopt iSA.

2.11 Glossary

bag-of-words: analysis of a document disregarding the order of words.

corpus: a set of documents consisting of texts.

followers: on Twitter, the other accounts who receive an update by a given one.

hashtag: (on SNS such as Twitter or Facebook) a word or phrase preceded by a hash or pound sign (#) and used to identify messages on a specific topic.

iSA: integrated Sentiment Analysis, a statistical algorithm for the estimation of the aggregated distribution of opinions.

lemmatization: reduction to the root of each token via morphological analysis of the words.

opinion: the motivation behind a sentiment: why positive? why negative?

sentiment: refers to the polarity of a comment/opinion, generally coded as positive/negative; pro/against, etc.

stemmization: reduction to the root of each token via removal of suffixes and prefixes.

tokenization: splitting a text into tokens (words), not trivial for Asian languages.

3

Extracting Subjective Well-Being from Textual Data

3.1 From SNS Data to Subjective Well-Being Indexes

In this chapter we will present some approaches to extracting aspects of well-being as expressed by users on social media networking sites (SNS). We will discuss four approaches: the *Hedonometer* proposed by Dodds, Harris, et al. (2011) in Section 3.2, the *Gross National Happiness Index* proposed by Rossouw and Greyling (2020) in Section 3.3, the World Well-Being Project in Section 3.4 developed by Jaidka et al. (2020) and our *Twitter Subjective Well-Being* index in Section 3.5. Although all these methods rely mainly on Twitter as data source, the same ideas can be straightforwardly applied to other sources of textual data like, e.g., Facebook.

In all the applications presented, data have been collected through both the public *search* and *streaming* Twitter API or by making a commercial agreement with Twitter itself or its subsidiary data resellers. For research or pilot studies like ours, the usage of public Twitter API coupled with some efficient *divide and conquer* data collection strategy would be enough as we noticed that, comparing the volumes of the collected tweets containing some given keywords with the volumes obtained through a commercial provider, a very good congruity can be observed. In our case, due to intermittent lack of funds, the source data could not have been collected in some specific periods as it will be discussed in the following sections.

3.1.1 Pros & Cons of Twitter Data

It is important to remark that the Twitter posts we obtained do not belong to individuals randomly chosen from a physical population (Baker et al., 2013, Murphy et al., 2014). For what matters the machine learning sampling assumptions, the reference statistical population is the whole corpus of posts collected as explained in the above and the training set is just a sample from this reference population. In addition to that, it is well known that Twitter accounts cannot be uniquely associated to individuals and some accounts are more active than others. For these reasons, the focus of the analysis in Section 3.5 is on the total volume of the posts collected through the public

Twitter API's and all statistics are obtained at an aggregated and not individual tweet or account level. That is the reason why we think that the aggregated sentiment analysis algorithms described in Section 2.7 are best suited to construct indexes based on social media data. Nevertheless, for completeness, we present the applications of Sections 3.2, 3.3 and 3.4 that do perform individual classification of tweets and/or accounts.

Remind further that only data with a geographic information have been collected (1–5% of all Twitter posts). This further non-randomized selection of the data may introduce additional bias. To our experience in different analyses, what we observed is that if we take the analysis on the geo-localized tweets at province level and we aggregate the estimates at national level, we obtain similar results from the whole set of non-georeferenced tweets. From this personal and limited evidence, we can speculate, without any proof, that if this bias exists, it has a limited effect. Nevertheless, this is worth a further systematic investigation, though Chapter 4 presents some possible approach to take into account this geographical effect.

So these data clearly suffer from selection bias due to the fact that not all the people possess a Twitter account and those who have one, are not necessarily equally active, enable geo-tagging, etc. Another source of bias also potentially affects the analysis: Twitter platform does not return all data, but only a sample of it. Although Twitter itself does not disclose the sampling mechanism, this latter selection effect seems to be less hurting. Indeed, in a sequence of tests Hino and Fahey (2019) showed that data obtained through Twitter API's and those obtained through the complete Twitter fire-hose do not statistically diverge.

Again, an attempt to deal with selection bias will be presented in Chapter 4. In this chapter we only focus on how different authors have approached the problem of distilling information about well-being from the texts contained in tweets.

Despite all the above limitations, the advantage of using tweets is that the collection of data can be done in (almost) continuous time and in a wide range of subregional areas. Honestly, the volumes are so huge that it is almost impossible to doubt that some information exists in these data. The point is how to separate noise from signal. In Chapter 2, we have already discussed the advantages and limitations of the different approaches to text analysis and we assume that the reader is now familiar with the conclusions of Section 2.7.

Until the results of the analysis are aggregated and do not contain any personal data, the approach of using SNS data in subjective well-being research is probably the less intrusive among many traditional survey approach, as noticed by all the authors involved in this cross-discipline field (Dodds, Harris, et al., 2011, Rossouw and Greyling, 2020, Curini, Iacus, and Canova, 2015, Iacus, Porro, et al., 2015). Indeed, through the analysis of these social media data it would be clear that instead of asking something through a (web-based or traditional) survey, thanks to the human supervised analysis, or other machine learning techniques, it is possible to capture expressions of well-being from the texts directly.

3.2 The Hedonometer

The algorithm behind the Hedonometer (Dodds and Danforth, 2010, Dodds, Harris, et al., 2011) is similar in the spirit to WordScore of Section 2.4.1. Let T be a given text and let w_i be one word in a reference dictionary, each one with an average happiness score $h_{avg}(w_i)$ between 1 and 9, the weighted average level of happiness for the text T is computed as

$$h_{avg}(T) = \frac{\sum_{i=1}^{N} h_{avg}(w_i) f_i}{\sum_{i=1}^{N} f_i} = \sum_{i=1}^{N} h_{avg}(w_i) p_i,$$

where f_i is the (absolute) frequency of the i-th word w_i and p_i is its relative frequency. N is in principle the set of unique words, but as the final aim was to realize a real-time system, the set was fixed to $N = 50,000$ most frequent words in the overall Twitter corpus they collected.

In order to build the dictionary and attribute the average happiness score to the words, the authors used four different sources: Twitter, Google Books (English), music lyrics (1960 to 2007) and the New York Times (1987 to 2007). For each corpus, they compiled lists of words ordered by decreasing frequency of occurrence and then merged the top 5,000 words from each source, resulting in a composite set of 10,222 unique words.

Amazon's Mechanical Turk service,[1] which is a crowdsource way to perform human classification, was used to score the words on a nine-point scale of happiness: from 1 (sad) to 9 (happy). Table 3.1 is an extract from this reference dictionary. The word that has the highest happiness rank is *laughter* with an average score of 8.5 out of 9 while *terrorist* is the one with the lowest rank with a score of 1.3 out of 9. We do not discuss here about the quality of these scores but, as remarked also by Rossouw and Greyling (2020), "the Hedonometer cannot deal with the context in which words are used, as words in itself are evaluated and not the sentiment of the construct. For example, a phrase such as *I did not enjoy the holiday*, will attract a score of 7.66 for *enjoy* and 7.96 for *holiday*, thus reflecting an overwhelmingly positive sentiment, when actually the sentiment is negative." One should also argue the following: how would the words *pandemic* or *virus* (and *terrorist*) have changed their scores if the corpora and the classification would have been taken in 2020 rather than in 2011. What about the newly created word *COVID*? Indeed, the authors studied the appearance of the COVID-19 related n-grams on Twitter in several

---

[1] Amazon's Mechanical Turk service `https://www.mturk.com/`

Table 3.1
An extract of the rankings of the 10,222 words from the reference dictionary used to build the Hedonometer. Source: Dodds, Harris, et al. (2011).

Word	Happiness rank	Average happiness score
Laughter	1	8.50
Love	3	8.42
Rainbow	13	8.06
Congratulations	25	8.00
Enjoy	112	7.66
⋮	⋮	⋮
Ministry	4410	5.58
Arrested	10209	1.64
Deaths	10219	1.64
Terrorist	10222	1.30

languages (Alshaabi et al., 2020) and updated the dictionary and the score of the words over time.[2] In fact, as of 2020-03-28, the `labMT-en-v2` version of the dictionary, contains *terrorist* in position 10,169 over 10,186 still with a score of 1.3, but *coronavirus* in position 10,171 with score 1.34, *pandemic* in position 10,180 with score 1.6 as well as *covid* with rank 10,172. The dictionaries in English, German, Korean, Spanish, Russian, Chinese, Arabic, Portuguese and French are available at `https://hedonometer.org` along with all the live dashboards, papers and case studies.

3.3 The Gross National Happiness Index

The lexicographic approach has been proposed by Greyling, Rossouw, and Adhikari (2020), Rossouw and Greyling (2020) to derive their Gross National Happiness (GNH) Index. The authors use a two-dimensional approach to classify each individual tweet. They first classify into a 0-10 scale the texts from unhappy = 0 to happy =10, 5 being neutral or "neither happy or unhappy", then they cross this sentiment scale with emotions. In particular, eight emotional categories are considered: namely joy, anticipation, trust, disgust, anger, surprise, fear and sadness, based on Robert Plutchik' s psychoevolutionary theory of emotion (Plutchik, 1980, 2001). Although the main references do not fully clarify the methodology, we infer from another application of the same authors (Steyn et al., 2020) these technical details. The sentiment is extracted using the `syuzhet` R package (Jockers, 2015), which is a project of

---

[2]To our knowledge the updates occurred in 2014 (version 1) and 2020 (version 2).

the Nebraska Literary Lab. According to the authors of `syuzhet`, the name comes from the Russian formalists Victor Shklovsky and Vladimir Propp, who divided narrative into two components, the *fabula* and the *syuzhet*. The latter refers to the technique of a narrative whereas *fabula* is the chronological order of events. The `syuzhet` method, therefore, is related to the way the elements of the story (fabula) are organized (syuzhet). In other words, the `syuzhet` package attempts to reveal the latent structure of narrative by means of sentiment analysis. Instead of detecting shifts in the topic or subject matter of the narrative, the package reveals the emotional shifts that serve as proxies for the narrative movement between conflict and conflict resolution. In practice, the method looks at the narrative pattern and assigns a score to each word in the context of the sentence. The sentiment is either the mean or the sum of the scores. It seems to work very nicely for complex texts and lengthy paragraphs, but for simple and short sentences. For instance, consider the sentence we have previously examined: *I did not enjoy the holiday*; in this case, the estimated sentiment is still positive. On the other hand, the method is able to classify a more articulated sentence like this: *Technical Damage After Trump Threatens Higher Tariffs* (see also Steyn et al., 2020), while basic lexicographic methods like Minqing and Bing (2004) assign a neutral score. At present, `syuzhet` works only with Latin characters sets.

Going back to the GNH Index, in the second step the emotions are extracted using the NRC Word-Emotion Association Lexicon method (Mohammad and Turney, 2013). Then the dimensions can be correlated in different ways, to see if a particular happy or unhappy day is connected to specific emotions, though the proper GNH Index consists in the first step only. The project, although originally applied to the cases of New Zealand, Australia and South Africa, seems to be running now also for United Kingdom, Spain, France, Germany, Italy, Belgium and the Netherlands. A live daily dashboard is available at `https://gnh.today`.

3.4   The World Well-Being Project

Schwartz, Sap, et al. (2016) focused on the individual message classification as well as user-level classification. This study focused on Facebook posts. They considered the *satisfaction with life* or SWL (Diener, Emmons, et al., 1985, Diener, Ng, et al., 2010) and *Positive Emotions, Engagement, Relationships, Meaning, and Accomplishment* or PERMA (Forgeard et al., 2010) scales approach. Using Amazon's Mechanical Turk, for each of the 10 PERMA components (five positive and five negative), turkers indicated the "extent to which a message expresses" the particular component on a scale that ranged from *none* = 0 to *very strongly* = 6. For SWL, turkers indicated their agreement that the message indicates life satisfaction (0 = strongly disagree, 3 = neutral,

6 = strongly agree). This training set was based on 5,100 public Facebook status updates, randomly selected from among 230 million public Facebook messages that contained at least 50% English words.

At user-level the data was acquired through a dedicated `MyPersonality Facebook App` from users who agreed to share their status updates for research purposes. The users were asked to answer to the SWL scale. This resulted in a dataset of 2,198 individuals, having collectively written 260,840 messages used only as a test set.

Each message was decomposed in unigrams and bigrams and then categorized according to a set of 2,000 topics extracted via LDA (see Section 2.4.4), as well as according to the lexicographic dictionaries from Hedonometer (see Section 3.2) and LIWC (Linguistic Inquiry and Word Count) from Pennebaker, Francis, and Booth (2001). All these features are used to predict the well-being of the users of the Facebook app from their data. The idea of the approach is to combine a psychological approach through well reasoned dictionaries and a data driven approach through the topics and understand their relationships.

This methodology has then be applied to Twitter data to produce maps of happiness of the United States (Jaidka et al., 2020). In their study, the authors took a sample of 1.53 billion geotagged English tweets over 1,208 US counties. They compared Twitter-based county-level estimates with well-being measurements provided by the Gallup Sharecare Well-Being Index survey through 1.73 million phone surveys and find that word-level methods alone yielded inconsistent county-level well-being measurements due to regional, cultural and socioeconomic differences in language use. However, removing the first three most frequent words led to notable improvements in well-being prediction showing a correlation with the Gallup data at up to $\rho = 0.64$.

Although not exclusively, the World Well-Being Project seems to collect mostly studies in the United States. The dedicated portal `https://wwbp.org` contains several case studies and reference papers.

3.5   The Twitter Subjective Well-Being Index

We have seen in Chapter 1 that there are many ways of capturing the latent variable that we refer to as *well-being*. As mentioned, in this application we adopt the scheme proposed by the New Economic Foundation think-tank for its Happy Planet Index (Foundation, 2012). The indicator that we are going to build will be called **SWB** and it will consists of eight of the overall dimensions defined in the Happy Planet Index. In particular, these dimensions of well-being concern three different areas: *personal* well-being, *social* well-being and well-being at *work*. More in detail:

Personal well-being

emotional well-being: the overall balance between the frequency of experiencing positive and negative emotions, with higher scores showing that positive feelings are felt more often than negative ones (`emo`);

satisfying life: having a positive assessment of one's life overall (`sat`);

vitality: having energy, feeling well-rested and healthy while also being physically active (`vit`);

resilience and self-esteem: a measure of individual psychological resources, of optimism and of the ability to deal with life stress (`res`);

positive functioning: feeling free to choose and having the opportunity to do it; being able to make use of personal skills while feeling absorbed and gratified in daily activities (`fun`);

Social well-being

trust and belonging: trusting other people, feeling treated fairly and respectfully while experiencing sentiments of belonging (`tru`);

relationships: the degree and quality of interactions in close relationships with family, friends and others who provide support (`rel`);

Well-being at work

quality of job: feeling satisfied with a job, experiencing satisfaction with work-life balance, evaluating the emotional experiences of work and work conditions (`wor`).

This categorization crosses and complements the SWL (satisfaction with life) and PERMA (Positive Emotions, Engagement, Relationships, Meaning, and Accomplishment) scales mentioned in the above.

This project focuses on two countries: Italy and Japan for several reasons. First is that the these projects are part of international collaborations between these two countries in the field of social media analysis, in particular the Universitiy of Milan and the University of Insubria for Italy and the Tokyo University and Waseda University for Japan. Second, because Japan and Japanese are not covered, at present, by any of the above mentioned methods it is an interesting comparison to do. Further, Italy is not part of the NEF survey yet and hence this country becomes relevant. The aim of the project is to expand the construction of the SWB indexes for several other European countries as well as the United States for comparison purposes with the other methods presented in the above.

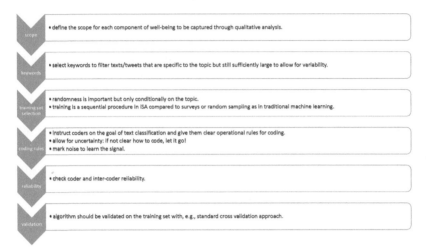

Figure 3.1
Workflow of analysis: from training set preparation to validation.

3.5.1 Qualitative Analysis of Texts

In our approach we use supervised sentiment analysis and, in particular, the iSA algorithm of Section 2.7. Being a supervised method, iSA requires a training set prepared by human coders as no dictionary or semantic rules will be used in this approach. This step of the process is crucial in that, being a qualitative analysis, all the relevant information is distilled at this stage but then it affects directly the quantitative part. Figure 3.1 is a summary of a typical workflow of the preparation of a training set. This workflow consists in these steps:

- what do we want to classify? i.e., satisfaction at work? health conditions?

- how do we reach the goal? Transform the goal into a set of operational rules for the human coders;

- which data have to included in the training set? Most of the time, this is not a random sample but a sequential sample obtained through the filtering of some keywords;

- was classification accurate? Check for inter-coder reliability;

- is the algorithm working properly? Do cross-validation on the training set.

The last step of cross-validation is as simple as follows: run the statistical analysis on the training set alone by splitting it into two random subsets. Use one of the two subset to predict the other subset and evaluate the quality of estimation in terms of discrepancy between the actual distribution and the estimated one[3] and replicate this analysis several times in order to evaluate an average quality of the statistical procedure. We don't show this step because in this chapter the focus is on explaining how to extract expressions of subjective (and perceived) well-being from textual data.

3.5.2 Data Filtering for Training-Set Construction

Let us start with an example. Suppose we want to classify tweets for the *emotional* component of well-being (`emo`). To this aim we select a random sample of tweets which contains at least one of the words in Table 3.2. For the other components of the **SWB** index, the filtering keywords are listed in Tables 3.3, 3.4, 3.5, 3.6, 3.7, 3.8 and 3.9. Remark that, even though the training set is built by filtering the data, the whole statistical analysis is done on the complete repository of tweets collected.

3.5.3 General Coding Rules

These rules apply to any sentiment analysis task.

- The first general rule is mark/tag/code `Off-Topic` posts appropriately. At this stage the machine learning algorithm will understand *noise*.

- The second rule is: if you are not fully convinced about the semantic context of a post do not classify it, just skip it go to the next one. These are not `Off-Topic`, let the algorithm try to classify it for you.

- It is admissible to classify `RT` (re-tweet) as original tweets from the account. This is an assumption of *transfer of emotion/opinion* which we assume to be the same as if they were expressed by the user account directly. Other researcher may disagree with this assumption of course.

- As each tweet can be classified along one or more dimensions, always try to consider parallel coding for all the categories and leave unanswered/ untagged those who do not apply to a given category.

3.5.4 Specific Coding Rules

For what concerns the application, the mandate for the coders was to consider only self-expressed or individual expression of well-being or own views of the tweeter. In order to obtain an index, the coders should classify the tweets in the categories: positive, neutral, negative and Off-Topic. Table 3.10 is an example of how texts can be distilled into emotions. Table 3.11 is another possible example.

---

[3]There exists several measures of discrepancy among distributions, like the Chi-squared or MAE indexes of formulas (2.5) and (2.4).

Unfortunately the natural language, or the real world in general, is more complex than the examples proposed in Tables 3.10 and 3.11. For example, the following real tweet (original Japanese version on the left, approximate English translation on the right):

体中が痛い…… 下手くそな証拠だ…… いくちゃんからメール来てたか ら元気でた よし、行動しよ	*My body hurts...* *Bad shit...* *I was fine because I received an* *email from Iku-chan.* *OK, act*

can be categorized under *resilience and self-esteem* (**res**) as *positive*, as well as this one:

意識高い高いと自尊心高い高いしてみたい

which translates approximately to

"High consciousness and high self-esteem".

More complex are tweets like this one:

精神とは控えめに見ても 90 パーセント妄想であって、妄想を自己と切り   離す作業を日夜続けることが、初期における人間の精神生活の主なノルマ   である。

which translates approximatively into

The spirit is 90% delusion, even if it is conservative, and continuing the   *work of separating the delusion from the self, day and night, is the main*   *norm of human mental life in the early days.*

The above text seems to express a negative view about life which we can arguably classify as *negative* for the component *satisfying life* (**sat**).

Plenty of examples can in fact be produced from real data. And tweets can be classified along one or more dimensions of interest. For example, the following one:

can be classified as *positive* for the components `emo` and `res` and *negative* for the component `vit`.

Table 3.2

Example of keywords used to select the training set data for the *emotional* (`emo`) component of the *personal* well-being. English is only shown as reference.

Italian	English	Japanese
Amore	Love	愛, 好き
Amicizia	Friendship	友情
Emozione	Emotion	感情
Sentimento	Sentiment	気持ち, 感じ
Felice	Happy	ハッピー, 喜, 嬉, うれしい
Felicità	Happiness	幸福, 幸せ
Lacrime	Tears	涙
Gioia	Joy	喜び
Divertente	Funny	おかしい
Triste	Sad	悲, 不幸
Depresso	Depressed	陰気, 愁い, 落ち込む, うつ病
Noia	Bored	退屈, うんざり

Table 3.3
Example of keywords used to select the training set data for the *satisfying life* (sat) component of the *personal* well-being. English is only shown as reference.

Italian	English	Japanese
Salute	Health	健康　調子　体調　元気
Malattia	Ill	病気
Famiglia	Family	家族
Figli	Children	子ども, 子供
Mamma	Mother	母
Papà	Father	父
Soldi	Money	金
Casa	Home	家

Table 3.4
Example of keywords used to select the training set data for the *vitality* (vit) component of the *personal* well-being. English is only shown as reference.

Italian	English	Japanese
Cinema	Cinema	シネマ
Teatro	Theater	劇場
Ristorante	Restaurant	レストラン -料亭
Palestra	Jim	ジム
Vacanza	Holidays	休日
Gita	Excursion	外出
Ferie	Holidays	休日
Pizza	Pizza	Pizza -ピザ
Fitness	Fitness	フィットネス
Jogging	Jogging	ジョギング
Tempo libero	Free time	レジャー
Volontariato	Voluntary	自主的な
Hobby	Hobby	趣味
Club	Club	クラブ
Circolo	Social club	社交クラブ
Stanco	Tired	疲

Table 3.5

Example of keywords used to select the training set data for the *resilience and self-esteem* (`res`) component of the *personal* well-being. English is only shown as reference.

Italian	English	Japanese
Fiducia	Confidence	信頼
Sicurezza	Safety	安全
Paura, Timore	Fear	恐怖
Capace	Capable	腕利き
Leader	Leader	棟梁
Ottimista	Optimistic	楽観的
Ottimismo	Optimism	楽観
Futuro	Future	将来
Fallimento	Failure	失敗
Obiettivo	Goal	ターゲット

Table 3.6

Example of keywords used to select the training set data for the *positive functioning* (`fun`) component of the *personal* well-being. English is only shown as reference.

Italian	English	Japanese
Libertà	Freedom	自由
Autonomia	Autonomy	自治
Significato	Meaning	意味
Imparare	Learn	学ぶ

Table 3.7

Example of keywords used to select the training set data for the *trust and belonging* (`tru`) component of the *social* well-being. English is only shown as reference.

Italian	English	Japanese
Aiuto	Help	助けて
Vicini Casa	Neighbors	隣人 -助
Rispetto	Respect	尊敬

Table 3.8
Example of keywords used to select the training set data for the *relationship*
(`rel`) component of the *social* well-being. English is only shown as reference.

Italian	English	Japanese
Famiglia	Family	家族
Figli	Children	子供
Mamma	Mother	母
Papà	Father	父
Fratello	Brother	兄
Sorella	Sister	妹, 姉
Amici	Friends	友人, 友達
Marito	Husband	夫
Moglie	Wife	妻
Parenti	Relatives	親族, 身内
Solitudine	Loneliness	孤独

Table 3.9
Example of keywords used to select the training set data for the *quality of job*
(`wor`) component of the well-being *at work*. English is only shown as reference.

Italian	English	Japanese
Iavoro	Job, work	作業, 職, 仕事, 職業, 作業
Carriera	Career	キャリア, 経歴, 来歴, 閲歴
Collega	Colleague	同僚
Ufficio	Office	事務所
Tempo lavoro	Working time	労働時間
Stress	Stress	ストレス
Disoccupato	Unemployed	失業者
Disoccupazione	Unemployment	失業
Contratto lavoro	Job contract	雇用契約
Stipendio	Salary	給与
Merito	Merit	実力, 有功

Table 3.10

Example of classification rule from fictitious texts with the aim of classifying the *emotional* (`emo`) component of the *personal* well-being.

Example (En)	Example (JP)	Classification
How lucky I am !	ラッキだ！	Positive
What a beautiful day :)	美しく晴れ渡った日	Positive
Fnally I passed the exam!	やっと合格した。	Positive
There are good and bad people	いい人と悪い人がいる。	Neutral
Tonight I have a date with my girlfriend <3	今晩彼女とデートする予定<3。	Positive
My girlfriend quit me last night	昨晩彼女に振られちゃった。	Negative
I feel sick and I have headache	風邪を引いて、頭が痛いんだ。	Negative

Table 3.11

Example of classification rule from fictitious texts with the aim of classifying the *trust and belonging* (`tru`) component of the *social* well-being.

Example (En)	Example (JP)	Classification
I was very happy that you gave me your support!	応援してくれてとても嬉しかった！	Positive
If only you work sincerely, you will be trusted.	誠実に働きさえすれば、あなたは信頼されるでしょう。	Neutral
It is by no means easy to satisfy everyone.	全ての者を満足させることをは決して容易ではない。	Neutral
I have no great belief in my doctor.	私は自分の医者をほとんど信頼していません。	Negative
She betrayed my trust.	彼女は自分の信頼を裏切った。	Negative

3.5.5 How to Construct the Index

Once the training set has been completely hand-coded, the iSA algorithm is applied to daily test sets of data. Each estimated distribution will contain the entries `positive`, `neutral`, `negative` and `Off-Topic`. The `Off-Topic` category represent the daily *noise* in the data, the rest represent the *signal*. For each component of the index, for example `emo`, the index is calculated, for day d and region r, as follows:

$$\texttt{emo}_{d,r} := \frac{\%\texttt{positive}}{\%\texttt{positive} + \%\texttt{negative}} \in [0,1]. \qquad (3.1)$$

The rationale behind formula (3.1) is that the intent is to capture expressions or judgments, and this is why comments not expressing any issue, judgments, etc., i.e. those classified as `neutral`, are removed from the calculation. The day d is easily extracted from the timestamp of Twitter data which is always present in the metadata of a tweet, while the region is obtained during the

Extracting Subjective Well-Being from Textual Data

Table 3.12
Yearly average values of SWB-I and SWB-J, their standard deviation in parentheses, and number of tweets in million. Data are in the period 01-02-2012/21-
06-2018 for Italy and 24-08-2015/31-12-2018 for Japan. For the Happy Planet
Index (HPI) and Human Development Index (HDI) source World Bank.

Year	2012	2013	2014	2015	2016	2017	2018
SWB-I	48.9	52.2	49.7	48.7	50.5	57.7	55.7
	(4.2)	(3.8)	(4.9)	(9.8)	(7.5)	(4.5)	(7.1)
Tweets	44.2 M	40.8 M	34.4 M	38.3 M	55.2 M	32.6 M	14.9 M
HPI	–	6.02	–	5.95	5.98	5.96	6.00
HDI	0.874	0.873	0.874	0.875	0.878	0.881	0.883
SWB-J	–	–	–	54.4	53.6	53.2	52.5
	–	–	–	(13.4)	(11.1)	(13.1)	(12.7)
Tweets	–	–	–	6.5 M	18.2 M	18.2 M	17.8 M
HPI	–	–	–	5.99	5.92	5.92	5.92
HDI	–	–	–	0.906	0.910	0.913	0.915

crawling. In our approach, to maximize the number of tweets form a given
region, we have crawl the data in different ways. Selecting data by keyword
and by country. In this case, if the tweet has specified latitude and longi-
tude in the meta data, we assign the tweet to the closest region on the map
by measuring the distance between the geographical center of a region and
the tweet's coordinates, otherwise we keep the tweet unassigned. A second
approach is to download the tweets geolocalized around a specific set of coordinates and a given radius. The third approach is to let Twitter identify the
location of a tweet. This is done by Twitter platform taking into account self-
declared information by the Twitter account and other metadata contained
in the tweet (if any). Then, at crawling stage, we also specify the name of the
region from which we want to crawl the data. Finally, duplicates obtained in
these different ways, are removed to come up with a clean data set.

3.5.6 The Data Collection

For the actual construction of the indexes SWB-I and SWB-J in this and the
next chapters, this project has collected tweets for both Italy and Japan along
different time spans. More precisely, tweets from those accounts which mainly
post in Italian (respectively Japanese) language or whose geo-reference information can be associated to Italy (respectively Japan). For Italy, the original
project collected about 250.4 millions of tweets in the period 01-02-2012/21-06-
2018, with a median of 50,000 tweets per day. For Japan, for several technical
reasons, we were able to collect at most 50,000 tweets a day, amounting to
about 60.8 millions of tweets in the period 24-08-2015/31-12-2018. Table 3.12

Table 3.13
Economic and environmental variables used in the econometric analysis of Section 3.7 plus two additional well-being indexes.

Variable	Frequency	Description
GDP growth	Quarterly	Percent change in quarterly real GDP year on year
Consumption growth	Quarterly	Percent change year on year
Investment growth GDP	Quarterly	Percent change year on year
Unemployment rate	Monthly	Percentage of work force
Life Expectancy at 40	Yearly	Male only
Happy Planet Index	Yearly	
Human Development Index	Yearly	

reports summary statistics. In the same table, we report also the values of the Happy Planet Index (HPI) by New Economics Foundation (2016) and the Human Development Index (HDI) by the United Nations Development Programme (2019), for the same years available for the SWB-I an SWB-J indexes. Data are taken from the data provider TheGlobalEconomy[4] as the other economic variables listed in Table 3.13, mostly coming from The World Bank, the International Monetary Fund, the United Nations, and the World Economic Forum. We also collected from OECD[5] the variable Life Expectancy of Males at 40, to capture the perception and quality of aging. This is the only yearly variable we consider as it varies non-monotonically in time and, moreover, is also correlated positively with economy growth in Italy ($\rho = 0.91$) but negatively in Japan ($\rho = -0.27$).

In this chapter we only consider national-level and low resolution statistics because of comparability between Italy and Japan as our collection of data do have different geographical and time resolutions. In Chapter 4 we will analyze regional aspects of the SWB-I index for which we have high resolution data, but for a shorter period than the data analyzed here. In Chapter 5 we will again compared Japan and Italy at higher time resolution focusing on the COVID-19 pandemic period and we also discuss the findings of different studies conducted through the alternative indexes presented in the above.

3.5.7 Some Cultural Elements of SNS Communication in Japan

As mentioned by Miyake (2007), in traditional Japanese communication, people tend to maintain distance and make sure and subjective experiences (Matsumoto, 1999). On textual analysis, most studies focus on the identification of big corpora of web blogs or SNS posts (Ptaszynski et al., 2014) in the context of sentiment and affect analysis. As emoticons, or emoji, are peculiar in Japanese written digital communication. For example, before the graphical emoticons appears, while in the western cultures horizontal emotions like " :) "

---

[4]TheGlobalEconomy.com
[5]https://data.oecd.org/healthstat/

were in use, in Japan (an other Asian countries) emoticons were and are traditionally vertical, like "(ˋoˊ)". Emoji's were already installed as a standard package of messaging platforms of mobile devices in Japan in the late 1990's (e.g., Jphone and iMode). The development of emoji is distinctive in Japan and arguably originates from the "kanji" culture where characters represent an idea or concept as a graphic symbol (which also applies to Asian countries that employ Chinese characters). Despite the abundance of emoticons, a large cross-country study seems to prove that regardless of the culture, vertical and horizontal emoticons convey similar concepts (Park, Baek, and Cha, 2014). Still, emoticons coupled with adverbs seem to be able to predict better than simple emoticons the affective perception of a text message (Rzepka et al., 2016). Many other studies related to the association between emoticons and emotions can be found in the literature (e.g., Shoeb and Melo, 2020, Novak et al., 2015) but they are not specific to the Japanese language. In relation to the Japanese culture to express emotions by non-verbal means, a graphic design known as the ASCII art has been quite extensively used in a bulletin board such as 2channel (one of the most popular online bulletin boards). Personality trait estimation of Japanese Twitter accounts have been studied in Kamijo, Nasukawa, and Kitamura (2016). Large scale studies on automatic sentiment tagging in Japanese during crisis periods can be found in (Vo and Collier, 2013). More linguistic analysis on specific Japanese terms related to likeness and happiness have recently appeared (for the word kawaii see, e.g., Iio, 2020) as well as gender-specific language studies (Carpi and Iacus, 2020).

Further note that, in discussing the concept of well-being in the Japanese society, Kumano (2018) distinguishes the two types of well-being: the *shiawase* or hedonic well-being, and *ikigai* or eudaimonic well-being. Clearly, both are captured in our analysis but not distinguished as this will be the scope of future work.

As for Italy, not so many equivalent studies have conducted apart the iHappy project by some of the same authors (Curini, Iacus, and Canova, 2015) which also has a counterpart project iGenki Japanese version still unreleased.

3.6 Preliminary Analysis of the SWB-I & SWB-J Indexes

Table 3.12 shows the yearly average values of SWB-I and SWB-J since 2012 for Italy, and since 2015 for Japan. What we notice is that the Japanese indicator shows a high medium-run stability in the range [52.5, 54.5]. In the Italian case, on the contrary, the medium-run variability is higher - the range of yearly values is [48.7, 57.7] and the value of SWB-I is, in some years, significantly lower or significantly higher than SWB-J. On the other hand, the standard deviations of the two indicators suggest (and the inspection of Figure 3.2

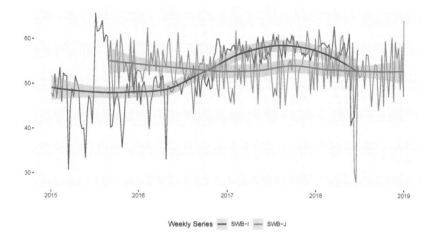

Figure 3.2
SWB-I and SWB-J weekly average series with estimated local-linear regression trends and standard errors bands. The peak in June–September 2015 for Italy is in correspondence of the Expo 2015 event.

proves) that the short-run volatility of SWB-J is definitely higher, compared to SWB-I. These results depict Japanese Twitter users as more reactive to the day-by-day events and emotions, while their evaluation of quality life is, on average, stable around quite satisfactory values. To be more exact, the Italian indicator show a high variability period, which basically coincides with the year when Italy organized and hosted the Expo 2015 event: that was an abnormal time frame, when heavily negative (due to controversies over delays in the preparation of the event and to allegation of bribery to some of the organizers) and strongly positive feelings (due to the appreciation and success of the event) rapidly emerged and changed in the Italian public opinion.[6]

An examination of the single components of SWB-J in Table 3.14 confirms the stability - along with a slight decline - of Japanese subjective well-being. A first paradox emerges looking at the well-being perception about social relationships: the outstanding value of the sub-component evaluating the quality of family relationship and friendship (`rel`) is, to some extent, at odds with the perceived well-being in terms of trust and sentiment of belonging (`tru`): this may indicate that the positive feelings nourished toward family and friends are not generalized to the rest of the society. The emotional sub-component of SWB-J is consistent with the global indicator, as the other dimensions related to personal well-being: slightly above the average we find the `fun` component, regarding the opportunity to do and to choose, and the involvement and

---

[6]In fact, despite several positive spikes, the average SWB-I in 2015 is the lowest of the examined period.

Table 3.14

The values of each components of the SWB-J index.

Year	SWB-J	emo	fun	rel	res	sat	tru	vit	wor
2015	54.4	54.8	59.3	75.2	54.4	56.9	35.4	43.2	55.9
2016	53.6	53.5	59.4	73.9	58.9	53.0	35.6	42.6	52.2
2017	53.2	51.0	57.9	75.7	55.9	51.5	36.1	42.7	55.0
2018	52.5	51.5	57.0	72.3	54.9	53.4	35.6	43.3	52.2

Table 3.15

The values of each components of the SWB-I index.

Year	SWB-I	emo	fun	rel	res	sat	tru	vit	wor
2012	48.9	60.5	67.8	34.1	55.1	43.9	59.2	53.9	16.4
2013	52.2	57.3	73.3	37.4	57.2	55.0	64.0	58.0	15.5
2014	49.7	48.2	68.3	39.7	56.1	52.4	62.6	55.2	15.1
2015	48.7	53.1	52.7	57.7	55.4	33.2	37.7	57.0	42.8
2016	50.5	62.2	40.5	65.9	59.7	30.2	28.9	58.4	58.0
2017	57.7	23.5	59.1	64.4	45.8	79.0	20.2	80.6	88.9
2018	55.7	40.4	57.8	59.1	46.4	64.9	26.6	74.5	76.2

satisfaction in daily activity; slightly below the self-perception of health and physical vitality (`vit`). Also the subjective well-being at work (`wor`) is strictly in line with the average SWB-J: this is in contrast with the Italian case (see Table 3.15), where the `wor` component is the most volatile and does not show strong correlation with the overall value of SWB-I, and likely documents the strong identification. Japanese people feel between their satisfaction as workers and their global well-being or that concerns at work are not expressed for cultural reasons which is not unexpected.

A look at the correlation between SWB-J (and SWB-I) and two well-known well-being indicators may raise some concerns. Table 3.16 shows a high correlation of SWB-J with the Happy Planet Index (HPI), developed for the first time in 2006 by the New Economics Foundation (2016). The HPI aims at giving a measure of sustainable well-being: it compares how efficiently residents of different countries are using natural resources to achieve long, high well-being lives. On the other hand, SWB-J is negatively related (with

Table 3.16

Correlation between the yearly average SWB-I and SWB-J and the two indexes Happy Planet Index and Human Development Index.

	Happy Planet Index	Human Development Index
SWB-I	0.14	0.80
SWB-J	0.81	−0.99

a correlation index equal to −0.99) to the Human Development Index, elaborated since 1990 by the United Nations Development Programme (UNDP), according to Amartya Sen's capability approach to well-being definition and evaluation (Robeyns, 2006). In measuring well-being, HDI takes into account three dimensions: health, education and material standards of living. It can be noted that the Italian SWB-I is positively related to both the indicators: a weak relation is shown with HPI and a strong one with HDI. All this should remind us, once again, that the plethora of well-being indices currently available seldom gives a measure of the same variable: each indicator addresses a specific definition of well-being and the relationships among all these definition are sometimes unclear and ambiguous. This does not imply that the measures provided are wrong or unreliable: it only require extreme clearness in explaining the methodology followed to construct the indicator, the data source, the definition of well-being the indicator aspires to account for.

3.7 Cross-Country Analysis 2015–2018 with Structural Equation Modeling

In this section we focus the attention on the impact of different economic variables on the SWB indicators. We make use of monthly and quarterly data of Table 3.13 and interpolate quarterly data at monthly frequency to make use of as much data as possible. Note that, while Italy is examined over the period 2012–2015, the analysis is restricted to the period 2015-2018 when both countries are considered together for comparison purposes. For the analysis we used the Structural Equation Modeling (SEM) with continuous response variable (Bollen, 1989) approach. SEM is a common method to test complex relationships between dependent variables, independent variables, mediators and latent dimensions. We assume that the true well-being is a latent variable influenced by the economic status of the country, which itself is supposed to be a latent variable, and by the health status of a country measured through the expectancy of life, and that the Twitter SWB-I/SWB-J indexes are observable measures of some aspects of the well-being latent variable.

In statistical terms, SEM consists of regression analysis, factor analysis, and path analysis to explore interrelationships between variables. It is a confirmatory technique where an analyst tests a model to check consistency between the relationships put in place. An alternative approach based on Bayesian Network Analysis (Pearl, 1995, Pearl and Russell, 2003) has been proposed in Cugnata, Salini, and Siletti (2020) for the province level Italian data used in Chapter 4. The following latent dimensions are theorized:

- Economy: captured by GDP growth, consumption growth, investment growth and unemployment rate;

- Well-being: we assume it is affected by the Economy latent variable and by the life expectancy taken as a proxy of the health conditions, and in turn the Well-being variable determines SBW-I/SWB-J measures. Then, a path diagram is constructed to represent inter-dependencies of the independent variables (GDP growth, consumption growth, investment growth, unemployment rate, Life Expectancy at 40), the latent dimensions and the dependent variable (SWB-I/SWB-J):

$$\text{Economy} \mapsto \text{GDP growth} + \text{Consumption growth}$$
$$+ \text{Investment growth} + \text{Unemployment rate}$$
$$\text{Well-being} \leftarrow \text{Economy} + \text{Life Expectancy at 40}$$
$$\text{SWB-J/SWB-J} \leftarrow \text{Well-being}$$

Further, the residual correlation among the observed variables is also captured in the model among the variable GDP growth and Life Expectancy at 40, Consumption growth, Investment growth and Unemployment rate. The results of the fitted model are presented in Table 3.17, while Figures 3.3–3.4 give a graphical representation of the same fitting. The models have been fitted using the `lavaan` package (Rosseel, 2012) and plots have been generated through the `semPlot` package (Epskamp, 2019).

3.7.1   Interpretation of the Structural Equation Model

In the Italian case (see Figure 3.4 and Table 3.17 bottom panel), all the observed economic variables have an expected and significant relationship with the Economy latent variable which, in turn, positively and significantly affect Well-being. A few anomalies, on the contrary, may be noted in the analysis of Japan. First of all, we notice from Figure 3.3 and Table 3.17 (top panel) that the relationship between the observable economic variables and the Economy latent variable, as well as the inter-dependencies among the observable economic variables, have the expected sign. Moreover, both the investment growth and the consumption growth rate show a significant relationship, whose coefficients are higher compared to Italy: being investment and consumption the main components of the aggregate demand, this likely explains why the relationship with GDP growth comes out to be statistically non-significant. But, on the other side, the unemployment rate is negatively and significantly related to the state of the Economy. Nevertheless, the Economy latent variable does not significantly affect Well-being: this is probably our main result and suggests that well-being perception among the Japanese is not determined by the objective, observable and mainly economic variables that have been traditionally used to measure the welfare of a country. It is worth reminding here that Diener, Suh, Smith, et al. (1995) observed the tendency of Asian cultures, compared to continental European and Anglo-Saxon ones, to mark a lower score in reported subjective well-being, economic conditions being similar, documenting probably that economic wealth is not necessarily the most important component of perceived well-being.

Table 3.17

Estimated coefficients for the SEM model applied to the Japanese (top) and Italian (bottom) data for the period 2015–2018.

		Relationship	Coefficient	Std. Err.
Japan 2015–2018				
Well-being	$\mapsto$	SWB-J	0.940***	0.101
Economy	$\mapsto$	Economic growth	0.406	0.497
Economy	$\mapsto$	Unemployment rate	−0.377**	0.148
Economy	$\mapsto$	Consumption growth	1.173***	0.159
Economy	$\mapsto$	Investment growth	0.730***	0.155
Well-being	$\hookleftarrow$	Economy	0.178	0.123
Well-being	$\hookleftarrow$	Life expectation at 40	−0.362**	0.159
Economic growth	cov	Life expectation at 40	−0.743***	0.174
Economic growth	cov	Consumption growth	0.404	0.525
Economic growth	cov	Investment growth	0.597*	0.358
Economic growth	cov	Unemployment rate	−0.440**	0.195
Italy 2015–2018				
Well-being	$\mapsto$	SWB-I	0.597***	0.113
Economy	$\mapsto$	Economic growth	0.514***	0.190
Economy	$\mapsto$	Unemployment rate	−0.581***	0.178
Economy	$\mapsto$	Consumption growth	0.597***	0.178
Economy	$\mapsto$	Investment growth	0.398**	0.179
Well-being	$\hookleftarrow$	Economy	0.921**	0.375
Well-being	$\hookleftarrow$	Life expectation at 40	0.834***	0.242
Economic.growth	cov	Life expectation at 40	0.246**	0.123
Economic.growth	cov	Consumption growth	0.121	0.137
Economic.growth	cov	Investment growth	0.230	0.134*
Economic.growth	cov	Unemployment rate	0.004	0.121

Note: *$p<0.1$; **$p<0.05$; ***$p<0.01$

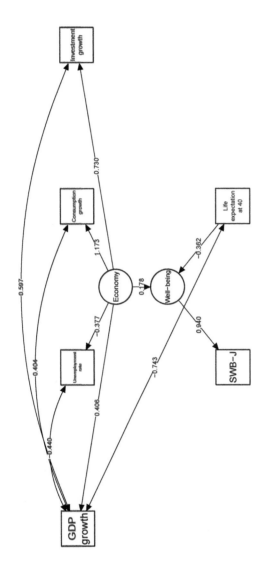

Figure 3.3
Graphical representation of the estimated SEM model for the Japanese data for the period 2015–2018.

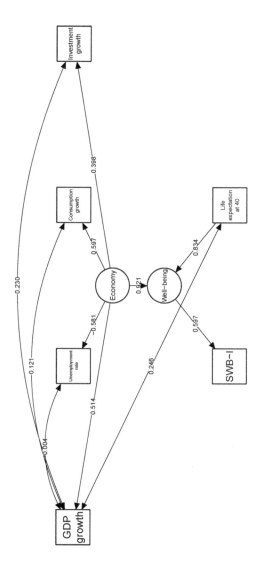

Figure 3.4
Graphical representation of the estimated SEM model for the Italian data for the period 2015–2018.

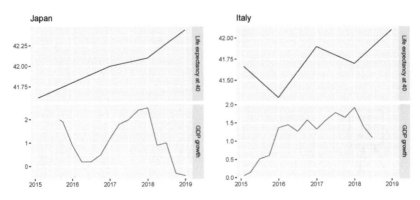

Figure 3.5
GDP growth and Life expectancy at 40 for Japan and Italy, period 2015–2018.

Life expectancy at 40, which we adopt as a proxy of public health conditions, shows an opposite relationship in Japan and in Italy both with the Well-being latent variable and the GDP growth rate. As shown in Figure 3.5, Life expectancy is increasing in both countries over the period 2015–2108. Similarly, the Italian GDP growth rate has a positive trend in these years, while the Japanese one is more fluctuating: this justifies the observed different sign of the relationship.

Lastly, the SWB indicator is positively and significantly related to Well-being in both countries. Nevertheless, the coefficient is higher in the Japan case, highlighting that SWB-J more closely resembles the well-being level, as depicted by the latent variable.

The structural equation model helps emphasizing differences between Japan and Italy, particularly with regard to the relevance of objective elements and circumstances in determining perceived well-being: in fact, it seems that Japanese subjective well-being is less affected by the economic conditions, suggesting that personal, relational and familiar dimensions may play a more important role in the individual and the social well-being sphere.

3.8 Glossary

API: Application Programming Interface, a server to server or server to consumer protocol to exchange data.

Mechanical Turk: Amazon Mechanical Turk is a crowdsourcing marketplace that allows for outsourcing processes and jobs to a distributed workforce who can perform these tasks virtually. More at https://www.mturk.com/

PERMA: Positive Emotions, Engagement, Relationships, Meaning, and Accomplishment (Forgeard et al., 2010).

SEM: Structural Equation Model.

SWL: Satisfaction with life (Diener, Emmons, et al., 1985, Diener, Ng, et al., 2010).

Twitter search API: API that gives access to a sample of tweets posted on a given day and satisfying a certain query.

Twitter streaming API: access point that gives access to a continuous stream of sampled data for a given search query.

4

How to Control for Bias in Social Media

4.1 Representativeness and Selection Bias of Social Media

As already mentioned extensively in Chapter 1 and in the different applications presented throughout the book, users accounts or data coming from SNS platforms are not a representative sample of the population of a country (see also Fan, Han, and Liu, 2014, Morstatter and Liu, 2017, Cesare et al., 2018, Hargittai, 2020, Wang, Yu, et al., 2020): they just represent the population of the users of those particular social media. Therefore, any well-being evaluation achievable from the analysis of these data cannot be immediately extended to the whole population.

Adjusting procedures can be applied to take into account for this bias within each model. Consider, as an example, Spyratos, Vespe, Natale, Iacus, et al. (2020): the authors explore the traveling behavior of migrant groups around the world using Facebook audience estimates and finding that reduced geographical mobility is associated with increased risk of social exclusion and reduced socio-economic and psychological well-being (Stanley et al., 2011, Vella-Brodrick and Stanley, 2013). In this study, there are two types of bias. Firstly, Facebook audience estimates represent the target marketing population and it is given in amount of MAU (monthly average users) subdivided by gender, age, location and a mix of interests. These data are affected by a thresholding and rounding bias (which, luckily do not affect Twitter data). Indeed, values of MAU below 1,000 are returned as 1,000 for confidentiality reasons and they are further rounded, with a rounding error that depends on the level of MAU: for example, for MAU values between 1,000 and 10,000, the rounding precision is 100; for values between 10,000 and 100,000, the rounding precision is 1,000; and so forth. This type of bias can be addressed by collecting data and aggregating the data at different resolutions, so it is less problematic. Secondly, there is the selection bias, which is the target of this chapter as it affects all social media. One possibility is to introduce the penetration rate as a proxy of the representativeness of the social media. Indeed, the authors define this penetration rate at country level as follows:

$$\text{pen\_rate}_t(age, gender, country) = \frac{\text{MAU}_t(age, gender, country)}{\widehat{\text{Pop}}(age, gender, country)},$$

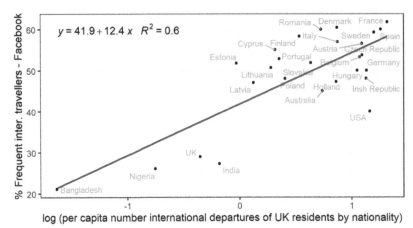

Figure 4.1
International traveling behavior in the UK by nationality/previous residence.
The x-axis shows the log of the per capita number of international departures
of UK residents by nationality from March 2018 to August 2018 as estimated
by the UK IPS survey. The y-axis shows the percentage of Facebook users who
live in the UK and have made at least one international travel from March
2019 to August 2019 by country of previous residence. Source: Figure 1 in
Spyratos, Vespe, Natale, Iacus, et al. (2020).

where $\widehat{\text{Pop}}(age, gender, country)$ are the UNDESA (2017) population projec-
tions by class of age, gender and country under the medium projection variant assumption for the year 2019, while $\text{MAU}_t(age, gender, country)$ are the
monthly average Facebook users in month t, also by class of age, gender and
country. After adjusting for the penetration rate, the Facebook derived travelling estimates are compared, at least for the UK departures, with the IPS
(International Passenger Survey) data (National Statistics, 2019) divided by
the estimated number of international departures by the stock of UK migrants by citizenship provided by Eurostat (2018). It turns out that these
data correlates quite well as shown in Figure 4.1.

A similar approach was used, for example, in Zagheni, Weber, and Gummadi (2017), Dubois et al. (2018), Spyratos, Vespe, Natale, Weber, et al.
(2019), Alexander, Polimis, and Zagheni (2020) and is related to the weighting strategy presented in Section 4.2.1. Other authors have tried different
approaches like, for example, a set of re-sampling strategies to make the social
media data more similar to a random sample, but these approaches require further assumptions (Wang, Yu, et al., 2020) or ground truth knowledge (Filho,
Almeida, and Pappa, 2015). Cugnata, Salini, and Siletti (2020) proposed to
mix social media data and survey data within the framework of Bayesian
Network Analysis (Pearl, 1995, Pearl and Russell, 2003) to extract derived
measures of well-being.

In this chapter, we will present a more systematic approach to the selection bias problem: our proposal[1] is to adjust statistics based on SNS data by anchoring them to reliable official statistics through a weighted, space-time, small area estimation model. As a by product, the proposed method also stabilizes the social media indicators, which is a welcome property required for official statistics. The proposed method can be used when official statistics exist at proper level of spatial resolution and when the social media penetration rate is known. As an example, we adjust a subjective well-being indicator of "working conditions" in Italy, and combine it with relevant official statistics. The weights depend on broadband coverage and Twitter rate at province level, while the analysis is performed at regional level. The resulting statistics are then compared to survey statistics on the "quality of job" at macro-economic regional level, showing evidence of similar paths.

4.2 Small Area Estimation Method

Small Area Estimation (SAE) models play an important role in sampling theory and are employed when one needs to produce estimates in areas that are smaller than those for which the survey was planned. A *direct* estimator $(\hat{y}_d)$, based only on the data coming from a limited-size sample from the small area, might be very unreliable; SAE *indirect* estimators are traditionally used to overcome this issue. Among indirect estimators, the so-called model-based estimators are obtained by an explicit regression model, where a relationship between the target variable and some covariates is assumed. Model-based estimators can be classified into *unit-level* models, when covariates are available at the unit level, and *area-level* models, when data are available only as area aggregates. In our case, as SWB-I and official statistics exist only at province or regional level, the only option available is the area-level model.

The basic area-level model is the Fay-Herriot (FH) model (Fay and Herriot, 1979), which is obtained as a linear mixed model in two stages consisting of a "sampling model" and a "linking model". Let $\hat{y}_d$ be a direct estimator of μ_d, a target unknown measure in area $d = 1, \ldots, D$: in the first stage, the "sampling model" (4.1) represents the uncertainty due to the fact that the target measure μ_d is unobservable and instead of it, only its measure on the sample $\hat{y}_d$ is available.

$$\hat{y}_d = \mu_d + e_d, \tag{4.1}$$

$\hat{y}_d$ is unbiased but unreliable due to the small observed sample; and the e_d are the sampling errors, which, given the characteristic of interest in d-th area, are

---

[1]Most of the content of this chapter, including figures and tables, is taken from Iacus, Porro, et al. (2020), published under the Creative Common licence.

assumed, for model convenience, to be independent and normally distributed (i.i.d.) with known variances, $N(0, \sigma_d^2)$.

In the second stage, the "linking model" (4.2), the area target measures μ_d are linearly related with a vector of area-level covariates $\boldsymbol{x}$.

$$\mu_d = \boldsymbol{x}_d'\boldsymbol{\beta} + u_d, \tag{4.2}$$

where $\boldsymbol{\beta}$ is the common regression coefficients vector, and the u_d are the model errors, unobserved and typically assumed i.i.d. from $N(0, \sigma_u^2)$. Combining the two model components (4.1) and (4.2), the final linear mixed model is defined as follow:

$$\hat{y}_d = \boldsymbol{x}_d'\boldsymbol{\beta} + u_d + e_d. \tag{4.3}$$

Several extensions of this basic area model have been proposed (Rao and Yu, 1994, Ghosh, Nangia, and Kim, 1996, Singh, Shukla, and Kundu, 2005, Marhuenda, Molina, and Morales, 2013) and recently, these models have also been used with big data (Porter et al., 2014, Marchetti, Giusti, Pratesi, et al., 2015, Marchetti, Giusti, and Pratesi, 2016, Falorsi et al., 2017) suggesting to use big data as covariates when official statistics are either missing or poor. Especially, big data are used as covariates in area-level FH models, because these data are often not available at the unit-level due to technical problems or legal restrictions, as is the case with social media search loads, remote sensing images or human mobility tracking.

Porter et al. (2014) used Google Trends searches as covariates in a spatial FH model, while in Falorsi et al. (2017), the time series query share coming from Google Trends was adopted as auxiliary variable to improve the SAE model-based estimates for regional Italian youth unemployment. Marchetti, Giusti, Pratesi, et al. (2015), Marchetti, Giusti, and Pratesi (2016) have shown that big data improve the precision of small area estimates when used along with traditional covariates (i.e., official statistics or administrative data). In detail, Marchetti, Giusti, Pratesi, et al. (2015) used big data as covariates in a FH model to estimate poverty indicators, accounting for the presence of measurement error, due to the availability of big data on mobility, following the Ybarra and Lohr (2008) approach. It is worth mentioning that Marchetti, Giusti, Pratesi, et al. (2015) suggested to make use of survey data in some way to take into account the selection bias caused by the use of big data, but they didn't pursue the goal. This work is an attempt to implement their idea in a systematic way.

Marchetti, Giusti, and Pratesi (2016) instead, used data coming from Twitter (Curini, Iacus, and Canova, 2015) as an instrumental covariate to estimate the Italian household share of food consumption expenditures at a provincial level, i.e., they exploit the correlation between the official statistics indicator and the social media data at regional level to reconstruct the official statistics at sub-regional level thanks to the granularity of the Twitter data.

Conversely to the scholars cited above, in our proposal, we do not use social media data (SWB-I) as covariate in a SAE model, but as direct measure of the target unknown variable (well-being), and adopting official statistics as covariates in the area model. Following this goal, because social media data are biased, before applying the model we endorse a weighting procedure as discussed in the next section.

4.2.1 Weighting Strategy

Usually, the methods adopted in the literature to face the selection bias problem when using non-representative samples (e.g., the propensity score weighting (Rosembaum and Rubin, 1983) or the Heckman correction (Heckman, 1979)) are based on the use of unit level data (Cooper and Greenaway, 2015). This also happens with social media data when individual characteristics of social media users are available, but in light of the recently established privacy rules (General Data Protection Regulation [GDPR]) this is an increasingly remote eventuality. Remind that, for Twitter data, the individual characteristics of every single account are not accurate or even unavailable and that SWB-I is calculated as an aggregated estimate at province level. Unfortunately, as we will see later on, as the official statistics are available only at regional level, we adopt a hierarchical aggregation of the data at regional level, weighted by the characteristics of provincial macro-variables. As it will be explained via an application in Section 4.3, the macro-variables consist of the broadband coverage and the Twitter rate at provincial level. The aim is to take into account the selection bias that comes from the fact that not all people use or can use Internet and, among those who use Internet, not all of them make use of Twitter. The Twitter rate also compensates for the difference in Twitter volumes that we observe through the different geographical areas.

In particular, in Section 4.3, we consider $\hat{y}_{dt}^w$ as the regional sampling mean, where the regional units are the weighted means of province level units, in order to overcome the non-random sampling structure of the data:

$$\hat{y}_{dt}^w = \frac{1}{\sum_{i=1}^{n_{dt}} w_{idt}} \sum_{i=1}^{n_{dt}} y_{idt} w_{idt}, \qquad (4.4)$$

where n_{dt} is the number of provinces in region d at time t, and w_{idt} are the weights. The choice of the actual weights depend on the application at hand. In Section 4.3 we will give a practical example. As an estimator of the variance of (4.4), we adopt the plug-in estimator for weighted means:

$$\sigma_{\hat{y}_{dt}^w}^2 = \frac{1}{n_{dt}} \left[\frac{1}{\sum_{i=1}^{n_{dt}} w_{idt}} \sum_{i=1}^{n_{dt}} y_{idt}^2 w_{idt} - (\hat{y}_{dt}^w)^2 \right]. \qquad (4.5)$$

4.2.2 The Space-Time SAE Model with Weights

Since SWB-I data are available for several periods of time T and domains D, we have chosen a particular SAE model, the spatio-temporal Fay-Herriot

(STFH) model, proposed by Marhuenda, Molina, and Morales (2013), to account for time and space correlations. This extension considers the spatial correlation between neighboring areas while simultaneously including random effects for the time periods nested within areas. Thus, for domains $d = 1, 2, \ldots, D$ and time periods $t = 1, 2, \ldots, T$, let μ_{dt} be the target unknown measure (well-being) in area d at time t. The STFH model, just as any FH model, is composed of two stages. In the first stage, the "sampling model" is defined as:

$$\hat{y}_{dt}^w = \mu_{dt} + e_{dt}, \quad e_{dt} \overset{ind}{\sim} N(0, \sigma_{\hat{y}_{dt}^w}^2), \quad d = 1, 2, \ldots, D, \quad t = 1, 2, \ldots T, \qquad (4.6)$$

where e_{dt} are the sampling errors that are assumed to be independent and normally distributed, and $\sigma_{\hat{y}_{dt}^w}^2$ is an estimator of the variance as defined in (4.5).

In the second stage of the STFH model, the "linking model" is as follows:

$$\mu_{dt} = x_{dt}'\beta + u_d + v_{dt}; \quad u_d \overset{ind}{\sim} N(0, \sigma_1^2); \quad v_{dt} \overset{ind}{\sim} N(0, \sigma_2^2), \qquad (4.7)$$

where x_{dt} is the column vector with the aggregated values of k covariates for the d-th area in t-th period of time and β is the vector of regression coefficients; u_d are the area effects that follow a first-order spatial autocorrelation process, SAR(1), with variance σ_1^2, spatial autocorrelation parameter ρ_1 and a $(d \times d)$ proximity matrix W. Specifically, W is a row-standardized matrix obtained from an initial proximity matrix W^I whose diagonal elements are equal to zero and residual entries are equal to one, when the two domains are neighbors, and zero otherwise. Normality of u_d is required for the mean squared error, but not for point estimation. Furthermore, v_{dt} represents the area-time random effects that are assumed i.i.d. for each area d; these effects follow a first-order autoregressive process, AR(1), with the autocorrelation parameter ρ_2 and variance equal to σ_2^2. Accordingly, the final proposed linear mixed model is:

$$\hat{y}_{dt}^w = x_{dt}'\beta + u_d + v_{dt} + e_{dt}. \qquad (4.8)$$

Therefore, $\theta = (\rho_1, \sigma_1^2, \rho_2, \sigma_2^2)$ is the vector of unknown parameters characterizing the spatio-temporal STFH model. Following Marhuenda, Molina, and Morales (2013), who provided $\hat{\beta}$, the empirical best linear unbiased estimator (EBLUE) of β, and $\hat{u}_d$ and $\hat{v}_{dt}$, the empirical best linear unbiased predictors (EBLUPs) of u_d and v_{dt}, both obtained by replacing a consistent estimator $\hat{\theta}$ in the respective BLUE and BLUPs introduced by Henderson (1975). The empirical estimation $\hat{\mu}_{dt}$ under the STFH model is given by:

$$\hat{\mu}_{dt} = x_{dt}'\hat{\beta} + \hat{u}_d + \hat{v}_{dt} \qquad (4.9)$$

As in Marhuenda, Molina, and Morales (2013), we use parametric bootstrap to estimate the mean squared error (MSE) of the EBLUPs. The MSE is calculated as follows:

$$MSE(\hat{\mu}_{dt}) = \frac{1}{B} \sum_{b=1}^{B} (\hat{\mu}_{dt}^b - \mu_{dt}^b)^2, \qquad (4.10)$$

where "b" remarks that these estimation is performed with the bootstrap procedure. And

$$\mu_{dt}^b = \boldsymbol{x}_{dt}' \hat{\boldsymbol{\beta}} + \hat{u}_d^b + \hat{v}_{dt}^b \tag{4.11}$$

is the empirical estimation obtained in the first step of the bootstrap procedure using the bootstrap area and time effects: $\hat{u}_d^b$ and $\hat{v}_{dt}^b$. In this way, the point estimates $\hat{\mu}_{dt}$ (indirect measure of well-being) of μ_{dt} (unknown well-being) can be supplemented with (4.10) as a measure of uncertainty.

4.3 An Application to the Study of Well-Being at Work

Having the opportunity of integrating existing information on well-being with more information, as those provided by social networks, as for SWB-I, with a strong subjective and perceived trait, is a very interesting goal. In this section with an application to Italian context we chose to use SWB-I index and official statistics to guide our proposal. Especially, in Section 4.3.1, we describe the data we use to implement the weighted procedure and the SAE model, and in Section 4.3.4 we discuss the result obtained.

4.3.1 Data and Variables

SWB-I index over the 24 quarters from 2012 to 2017 is available at provincial and regional level. Then the outcome variable is the estimated wor component of SWB-I index and this represents the input variable y_{idt} in formula (4.4).

As the variability of the number of tweets is remarkable, both along the time and the space dimension there is the need to take into account this diversity. The range of data extends from a minimum of 1,727 tweets in 2016-Q1 for the Basilicata region to a maximum of 2,728,640 in 2017-Q2 for the Lombardia region (Notice that Valle d'Aosta has been dropped from the analysis because, considering that it consists of a single province, the proposed approach is not applicable because, e.g., random effects cannot be estimated).

In order to have a more reliable view of the SWB-I data at the regional level, we use the *Twitter rate* (i.e., the ratio between the number of tweets analysed and the population size in the area in the same period). The distribution of the Twitter rate over time among the Italian regions is shown in Figure 4.2, the average Twitter rate is around 18% ($SD = 12.29$), with a minimum regional value higher than 9% ($SD = 4.93$) in Campania, and a maximum regional value higher than 30% ($SD = 21.15$) in Molise (time averages for all the regions are blue points in the figure). The dispersion during the observational period is lower for large regions like Lazio, Puglia, Campania and Lombardia, while being higher for small regions like Molise and Marche.

The better comprehension of the SWB-I information using the Twitter rate is made evident by examining Figure 4.3. The Twitter counts of 2017-Q4,

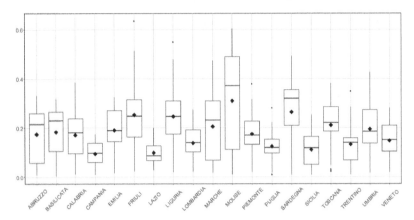

Figure 4.2
Twitter rate for the considered Italian regions from the first quarter of 2012
to the last quarter of 2017.

shown on the left of the figure, give the erroneous impression that most of
the SWB-I information comes from only a few large more populous regions
(Piemonte, Lombardia, Veneto, Emilia and Campania), while the Twitter rates
displayed on the right of the figure give the correct conclusion that all regions
are homogeneously monitored.

4.3.2   The Construction of the Weights

To implement the weighting procedure introduced in Section 4.2.1, after a
selection process to define significant variables, we use the Twitter rate and
the broadband coverage. Twitter rate is closely related to mobile phone shares
and broadband coverage is a measure of Internet capability. The usage of these
two variables is an attempt to take into account the selection bias. The Twitter
rate, computed in each period and at province level, can be considered a good
proxy of the use of Twitter for Italians. The broadband coverage is annual
public data provided by *Il Sole 24 Ore* and *Infratel Italia* for all the Italian
provinces and can be considered the opportunity to access the Internet in
the different provinces. Coverage is quite stationary during a single year but,
over time, what can happen is only an improvement of coverage in space or
in signal intensity. Therefore, we replace the missing values with the data
from the previous year to ensure that the coverage is not overestimated. The
average broadband coverage is around 94% ($SD = 4.68$), with a minimum
regional value of 72% ($SD = 4.57$) for Isernia in the Molise region. In 2012,
the coverage mean was 92.15% ($SD = 3.9$) and in 2017, it was 92.65% ($SD = 5.6$). So, during the examined time period, the average broadband coverage
remained quite the same, but the variability among regions increased, with an
increment of around 42%. In detail, calling $w_{1,idt}$ the Twitter rate and $w_{2,idt}$

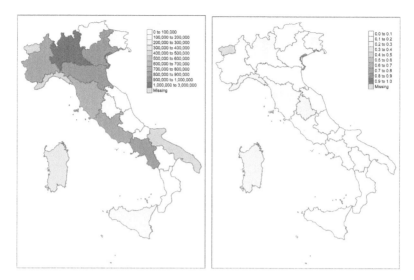

Figure 4.3
Twitter counts map, on the left, and Twitter rates map, on the right, in the last quarter of 2014.

the broadband coverage, to apply to weighting procedure for $\hat{y}_{dt}^{w}$ in (4.4) and for $\sigma_{\hat{y}_{dt}^{w}}^{2}$ in (4.5), we computed the weights as $w_{idt} = w_{1,idt} \cdot w_{2,idt}$.

4.3.3 Official Statistics to Anchor the Model

To apply the model proposed in Section 4.2.2, we need official statistics to use as covariates. After the Stiglitz's Commission suggestions, the Italian scenario of well-being measurement has increasingly changed. For example, the Italian National Institute of Statistics (ISTAT) set up the equitable and sustainable well-being project, where they plan a very complex system of well-being indicators, just following the same Commission suggestions. In 2013, they provided the BES ("Benessere Equo e Sostenibile", that in English is "Fair and Sustainable Well-Being") index for the Italian regions, which analyses several dimensions of well-being. Among these, the "work and life balance" dimension is the one that more closely relates to our research, though the construction of the composite indicator changed through time and it is not available for all quarters and provinces of Italy, making it impossible to use it in our study.

ISTAT also provides other measures of well-being from the sample survey "Aspect of daily life"; however, these indicators are annual and representative for the five Italian macro-economic areas: North-East, North-West, Center, South and Islands.

Dropping out the idea to use the BES indexes and the "Aspect of daily life" survey measures, as covariates, we decided to rely on the only official

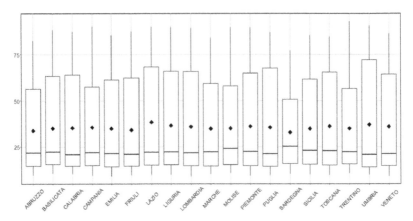

Figure 4.4
The SWB-I's unweighted wor dimension ($\hat{y}_{dt}$) for the considered Italian regions
from the first quarter of 2012 to the last quarter of 2017.

statistics distributed by ISTAT that are available at least at the regional level
and for the period of the analysis (Although only every quarter, the ISTAT
data are available: `http://dati.istat.it/` and `http://demo.istat.it/`).
Despite the fact that the proposed model should work for each component
of the SWB-I at the province level, due to the limited availability of official
statistics at frequency higher than the year and at the sub-national level,
we restrict our empirical analysis to the wor dimension of the SWB-I. Even
though the wor dimension could be monitored daily at province level, for the
analysis they have been aggregated quarterly for each province (y_{idt}).

The distribution of the unweighted wor ($\hat{y}_{dt}$) with regional aggregation over
time is shown in Figure 4.4. The average of wor is 35.34% ($SD = 25.40$) with
a minimum average regional value around the 33% ($SD = 21.01$) in Sardegna
and a maximum average regional value higher than 38% ($SD = 28.48$) in
Lazio. The minimum and the maximum value of the quality of work are 9.01%
for Lombardia in 2012-Q2 and 93.01% for Trentino in 2015-Q3, respectively,
with similar averages of 35.79% ($SD = 26.87$) and 34.88% ($SD = 24.74$),
respectively.

The considered area level auxiliary variables, before any process of selection, in the job context were as follows: the unemployment and inactivity
rates, computed both in relation to the labour force (as they are traditionally calculated) and to the resident population; and the birth, the mortality
and the natural rates, in the socio-demographic context. In the numerator of
the natural rate there is the natural balance, which is the difference between
births and deaths. After fitting the model, the selected covariates that make
up the matrix x in the model (4.8), are the "unemployment rate" x_1 and the

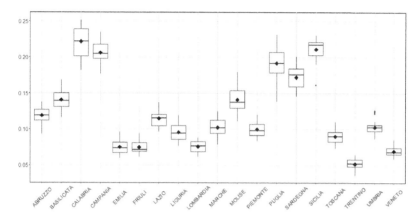

Figure 4.5
Unemployment rate (x_1) for the considered Italian regions from the first quarter of 2012 to the last quarter of 2017.

"mortality rate" x_2. The selection of these variables is the result of a standard model selection procedure after testing different variable configurations.

A large number of studies - since Clark and Oswald (1994) - provides documentary evidence of the negative relationship between unemployment and subjective well-being. It has also been argued that getting unemployed people back to work can do more for their well-being perception than subsidizing their unemployment status (see, e.g., Winkelmann (2014)). In other words, non-pecuniary costs of unemployment are significant: therefore, higher unemployment rate (i.e., a higher risk of being unemployed) is here assumed as related to the evaluation of well-being at work.

The relationship between working conditions and subjective well-being is often mediated, in the same literature, by health conditions: mortality or morbidity rates are assumed, in this respect, as proxies of health conditions.

The distribution of the unemployment rate over time among regions, as shown in Figure 4.5, reveals an average unemployment rate of 12.37% ($SD = 5.31$), with a minimum average regional value around 5% ($SD = 0.78$) for Trentino and a maximum average regional value higher than 22% ($SD = 2.13$) for Calabria. The same two regions also register the minimum and maximum values for the unemployment rate, 3.59% in 2017-Q3 and 25.15% in 2017-Q4, respectively.

The distribution of the mortality rate over time among regions, as shown in Figure 4.6, illustrates an average mortality rate of 0.267% ($SD = 0.04$) with a minimum average regional value around 0.216% ($SD = 0.022$) in Trentino and a maximum average regional value higher than 0.343% ($SD = 0.032$) in Liguria. The same two regions also register the minimum and maximum values for the mortality rate, 0.19% in 2014-Q3 and 0.42% in 2017-Q1, respectively.

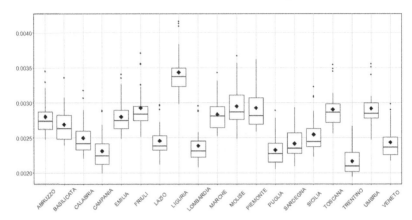

Figure 4.6
Mortality rate (x_2) for the considered Italian regions from the first quarter of
2012 to the last quarter of 2017.

4.3.4   Results of the SAE Model

The weighted quality of job dimension $\hat{y}_{dt}^w$ (weighted wor), obtained following
(4.4), has remained stable with little variability between regions (Figure 4.7).
The distributions were compressed until the second half of 2015, when it grew.
This is especially evident from the second half of 2016, when this dimension
attained values greater than 80, and even the differences between the regions
were more marked, the box-plots less crushed. Moreover, the average of $\hat{y}_{dt}^w$
is 36.17% ($SD = 26.38$) with a minimum average regional value around 34%
($SD = 22.91$) for Sardegna and a maximum average regional value higher
than 39% ($SD = 29.24$) for Lazio, reflecting the earlier distributions shown in
Figure 4.4 for $\hat{y}_{dt}$ (unweighted wor). The minimum and maximum values of
the $\hat{y}_{dt}^w$ remained with Lombardia in 2012-Q2 (8.99%) and Trentino in 2015-
Q3 (92.76%), respectively, and their averages were still similar (36.68% with
$SD = 27.46$ for Lombardia and 37.99% with $SD = 28.66$ for Trentino).

Since comparing rankings is a valuable tool for policy makers and analysts,
here we propose some discussions about them. The different rankings obtained
by the two indices, both unweighted $\hat{y}_{dt}$ and weighted $\hat{y}_{dt}^w$, show no differences
for around 4% of the cases (Δ = ranking differences), and only 15.6% of the
cases show a Δ greater than four positions. The mean of the Δ is equal to 2.19
($SD = 2.58$). Regions with the greatest differences were Trentino, Campania,
Marche, and Sardegna, with the first two showing position improvement and
the last two showing position weakening. For Trentino in particular, we remark that, after the weighting procedure, the greatest improvement took place
during all four quarters of 2017.

In the applied STFH model (4.8) data are available for $T = 24$ time
instances, and the domains are $D = 19$, the considered Italian regions. Our

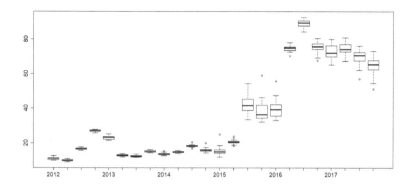

Figure 4.7
The SWB-I's weighted `wor` dimension $(\hat{y}_{dt}^w)$ during the periods from the first quarter of 2012 to the last quarter of 2017.

data are "balanced" in that each region is measured using the same number of times and on the same occasions.

The row-standardized proximity matrix $\boldsymbol{W}_c$ of dimension 19×19 has been obtained from an initial proximity matrix, $\boldsymbol{W}_c^I$, whose diagonal elements are equal to zero and residual entries are equal to one, when the two regions had some common borders, and zero otherwise. Since in Italy, there are two regions corresponding to two islands (Sicilia and Sardegna), for these regions, we take other Italian regions with direct naval connections as neighbors.

As shown in Table 4.1, the coefficients for the covariates ($\hat{\beta}_1$ and $\hat{\beta}_2$) were both negative. This means that regions with larger unemployment and mortality rates had a poorer quality of job dimension. The estimated spatial autocorrelation coefficient $\hat{\rho}_1$ is significant enough with a small negative value of about -0.07, (the size of the vector used is not large, $D = 19$), while the temporal autocorrelation coefficient $\hat{\rho}_2$ is still significant and has a greater positive value equal to about 0.88. The value equal to zero for $\hat{\sigma}_1^2$ is coherent with the analysis of distribution discussed above. The quality of job changes over time, but either little or not at all between regions.

4.3.5 A Weighted Measure of Well-Being at Work

In Figure 4.8, the scatter plots between the resulting $\hat{\mu}_{dt}$, obtained by fitting the STFH model, and the direct estimates, both unweighted $\hat{y}_{dt}$ (on the left) and weighted $\hat{y}_{dt}^w$ (on the right). In the SAE context, this graphical representation is used to test if the estimates are design unbiased: if the points lie along the diagonal, the direct estimates are approximately design unbiased, but if the points are under the line, the direct estimators are larger than the values predicted by the model, and vice versa if the points are above the line. Both

Table 4.1
STFH model results. Top: the estimated regression coefficients $\hat{\beta}$ in (4.9).
Bottom: the estimated values for the vector of predictors $\hat{\boldsymbol{\theta}}$.

$\hat{\beta}$	Estimate	Std. Error
Intercept	62.72***	5.49
Unemployment rate	−82.63***	31.11
Mortality rate	−5,649.48***	1,450.95

$\hat{\boldsymbol{\theta}}$	Estimate	Std. Error
$\hat{\sigma}_1^2$	0.00***	$\sim 1 \times 10^{-3}$
$\hat{\rho}_1$	−0.07***	$\sim 1 \times 10^{-3}$
$\hat{\sigma}_2^2$	94.72***	$\sim 1 \times 10^{-3}$
$\hat{\rho}_2$	0.89***	$\sim 1 \times 10^{-3}$

Note: $^*p < 0.1$; $^{**}p < 0.05$; $^{***}p < 0.01$

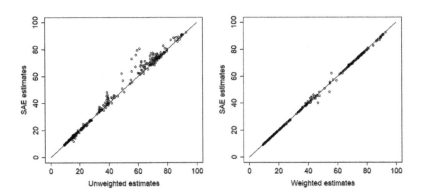

Figure 4.8
Predicted values from the STFH model $\hat{\mu}_{dt}$ (SAE estimates) vs estimates of
`wor`, unweighted $\hat{y}_{dt}$ on the left and weighted $\hat{y}_{dt}^w$ on the right.

the plots in the figure show points that lie along the diagonal for most of the cases. On the left side of the figure, we compare the SAE estimates $\hat{\mu}_{dt}$ with $\hat{y}_{dt}$, the unweighted estimates of wor, and there are more points away from the diagonal line than when the same estimated values are compared with $\hat{y}_{dt}^w$, the weighted estimates. Looking at the same plots, but for the different considered quarters, we find that the points away from the diagonal are in the periods where we have fewer analyzed tweets, and we observe an anomalous value of the variances. These two situations are caused by a lack of reliability in the information, but overall, we can conclude that the weighted estimates $\hat{y}_{dt}^w$ are approximately design unbiased.

4.3.6 The Estimated Measure of Well-Being at Work from the SAE Model

Considering the rankings, what changes if we use SAE model estimates instead of direct estimates, whether weighted or not?

Comparing the rankings obtained with the $\hat{y}_{dt}$ and those obtained with $\hat{\mu}_{dt}$, we find that in 29.2% of the cases, the position is the same, and in 15.8% of the cases, the Δ is greater than four. The mean of the ranking Δ is 2.16 ($SD = 2.58$). Equally, when we compared the above simple means $\hat{y}_{dt}$ with the weighted means $\hat{y}_{dt}^w$, regions with the greatest differences are Trentino, Campania, Marche and Sardegna, with the first two showing position improvement and the last two showing position weakening. For Trentino, there is a great improvement during all quarters of 2017. Comparing the rankings obtained with the weighted values $\hat{y}_{dt}$ and those obtained with models estimates $\hat{\mu}_{dt}$ shows a very different situation: in 84.9% of the cases, the positions are identical, with less than 1% of the cases having a Δ greater than four (just one case has a great ranking difference: Marche in 2015-Q3 with a lag equal to eight positions). The average of the Δ equals 0.2 ($SD = 0.6$), which means that moving to weighted estimates $\hat{y}_{dt}^w$ with model predictions $\hat{\mu}_{dt}$ provides estimates that rank the same.

In SAE literature (Molina and Marhuenda, 2015), coefficients of variations (CVs) are used traditionally to analyze the gain of efficiency for model estimates. While national statistical institutes are committed to publishing statistics with a high level of reliability, it is generally considered that estimates with CVs greater than 20% are not reliable. In Figures 4.9 and 4.10, the CVs of the three compared indices are shown, for the proposed final STFH model, CVs were obtained by using the bootstrap procedure for the MSE estimates (4.10). As is evident, in our application the CVs are always lower than 20%, except for fewer peaks. Especially, for most regions, the CVs are lower than 10% (Figure 4.10), while peak values are obtained in only a few quarters for 13 regions: Calabria, Campania, Emilia, Friuli, Lazio, Liguria, Marche, Molise, Piemonte, Lombardia, Sicilia, Toscana and Trentino. We stress that these high values of CVs are not stationary for these regions and it is clear that whenever we observe a peak of CVs, both the weighted indices and the model estimates

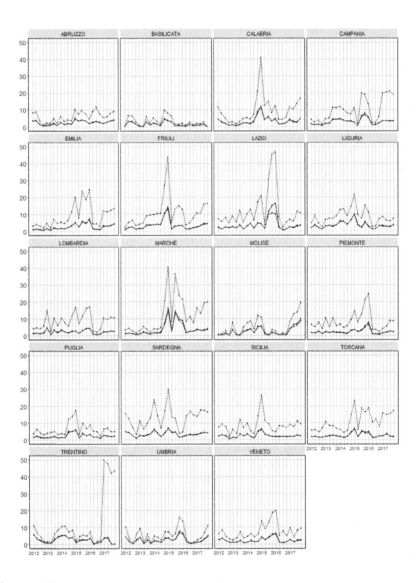

Figure 4.9
Coefficient of Variations for all the regions[2]; SAE estimates ($\hat{\mu}_{dt}$) with solid lines, weighted estimates ($\hat{y}_{dt}^{w}$) with dashed lines and unweighted estimates ($\hat{y}_{dt}$) with dotted lines.

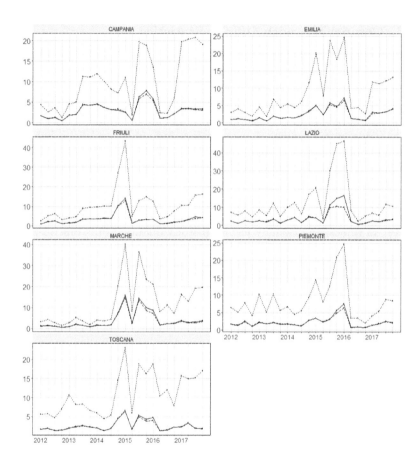

Figure 4.10
Coefficient of variations for the regions with peaks greater than $20\%^2$; SAE estimates ($\hat{\mu}_{dt}$) with solid lines, weighted estimates ($\hat{y}_{dt}^w$) with dashed lines and unweighted estimates ($\hat{y}_{dt}$) with dotted lines.

improve reliability. As well, CVs obtained for the model estimates ($\hat{\mu}_{dt}$, solid line) are always lower than the weighted estimations ($\hat{y}_{dt}^w$, dashed line) and the unweighted estimates ($\hat{y}_{dt}$, dotted line). (For model estimates are computed as $CV = 100 \times \frac{\sqrt{MSE}}{Index}$, while for the others are $CV = 100 \times \frac{\sqrt{Variance}}{Index}$). Thus, values based on a STFH model look less variable in terms of the CV.

4.3.7   Comparison with Official Statistics

In this section we compare our index obtained by the STFH model with an index of work satisfaction (WS) provided by ISTAT in its "Aspects of daily life" report.[2]

The ISTAT's sample survey "Aspects of daily life" is a part of an integrated system of social surveys - The Multipurpose Surveys on Household - and collects fundamental information on Italian individual and household daily life. It provides information on the citizens' habits and the problems they face in everyday life. In the questionnaire, there are several thematic areas, according to different social aspects, permitting to realize which is the quality of individuals life, the degree of satisfaction of their conditions, their economic situation, the area in which they live, and the functioning of all public utility services, all topics traditionally useful to study quality of life. Since 2005 this survey is annual with data collection in February.

For our purpose we only consider WS, defined as the percentage of employed persons, aged 15 years and over, with a "good" level of satisfaction with their work. This index is computed as the sum of the percentages of people declaring to be "quite" and "very much" satisfied during the survey. Yearly WS data are distributed free of charge, but, as mentioned previously in the covariates section, they are representatives for the five Italian geographical areas: North-west, North-east, Central, South and Islands.

To compare this index with our information, we aggregate the SAE estimates, $\hat{\mu}_{dt}$, obtained as discussed in the previous sections, yearly and in the same geographical areas, weighing with the corresponding resident population (SAE-wor).

The correlations between ISTAT index and SAE-wor are displayed in Table 4.2. If we consider all the overall data, the correlation is about 25%, while if we analyse the relationships within each area we find stronger links, with a maximum value in South Italy equal to 85%.

Given the different scales of the ISTAT index and the proposed STFH estimator, with the purpose of visual comparison, Figure 4.11 represents the plot of their values both standardized. Looking at these plots, the correlations become quite evident. We remark that the correlation results are similar if we replace the STFH estimator with the raw wor measures (unweighted $\hat{y}_{dt}^w$ and weighted $\hat{y}_{dt}^w$).

---

[2]All the details about the probability sample for the ISTAT survey "Aspects of daily life" can be found at https://www.istat.it/it/archivio/91926

Table 4.2

Pearson correlation coefficients ρ between ISTAT's WS and SAE-wor, in the five Italian geographical areas.

Area	Overall	North-west	North-east	Central	South	Islands
ρ	0.245	0.694	0.383	0.581	0.849	0.480

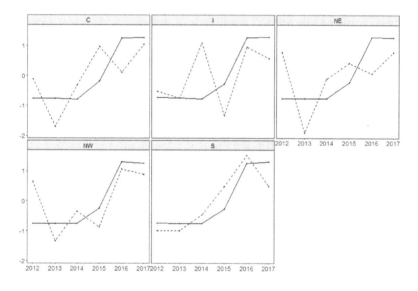

Figure 4.11

Standardized time series of SAE-wor, solid line, and ISTAT's WS, dotted line, in the five Italian geographical areas (C: Central, I: Islands, NE: North-east, NW: North-west, S: South).

4.4   Conclusions

Looking at the results, it seems that the selection bias inherent in social network data can be controlled using the spatio-temporal SAE approach. In detail, what we have shown is that - properly weighting statistics based on social media - we have approximately design unbiased statistics, i.e., we have corrected the selection bias up to the only benchmark data available, which are the official statistics. We also gained additional properties through the SAE model, one of which being the stabilization of the variances of the social media statistics, which is a property required by official statistics. We have also shown that, despite using SNS data, the adjusted "wor" component of SWB-I (though built upon different official statistics) correlates with the ISTAT statistics (available at macroeconomic level only) on the quality of work survey data.

This first attempt represents just a drop in the sea. Certainly, the accuracy of the proposed method could be improved using different SAE models based on dynamical systems, so to exploit fully the high resolution of the social media data, or by integrating more big data sources at the same time, each with its own bias corrected statistics. These kinds of extensions represent interesting methodological challenges for the future.

4.5   Glossary

Facebook Audience Estimates: Number of potential Facebook users with a given set of interest by gender, ages, location, etc. (Obtained through the Facebook Marketing Application Programming Interface (API)) (Facebook, 2020).

MAU: Monthly active users of the Facebook platform.

SAE: Small Area Estimation method.

5

Subjective Well-Being and the COVID-19 Pandemic

5.1 The Year 2020 and Well-Being

The aftermath of the worldwide COVID-19 outbreak is involving all the aspects of civil coexistence and modifying - we do not yet whether temporarily or permanently - our perception of life. This may result both from a more intense feeling of precariousness, from the restrictions imposed to our social interactions or from objective traces of impoverishment, due to the forced slowdown of economic activities.

Despite being the pandemic, by its nature, a global event, in this chapter we will focus mainly on the Italian and the Japanese case, though we report some examples from US and South Africa in the following. Plenty of studies have been carried out, in this short span of time, about the effect of pandemic on feeling, mood and health status - in particular, on mental health - both among Italian (Marazziti et al., 2020, Maugeri et al., 2020, Rossi et al., 2020, Sani et al., 2020, Gualano et al., 2020) and Japanese people (Yamamoto et al., 2020, Qian and Yahara, 2020, Ueda et al., 2020). Some of these studies, moreover, focus on specific population targets such as elderly and young people or unemployed workers (Orgilés et al., 2020, Shigemura, Ursano, et al., 2020). Particular attention is devoted to vulnerable categories and people involved in Covid-care activities: e.g., health care workers, who may suffer heavy emotional distress and even discrimination and stigmatization effects (Shigemura and Kurosawa, 2020, Asaoka et al., 2020, Torricelli, Poletti, and Raballo, 2020); pregnant women and newborns (Haruna and Nishi, 2020); patients with specific pathologies (Capuano et al., 2020).

All these studies aim at evaluating the impact of the pandemic on individual and collective well-being and suggesting intervention priorities.

On the other hand, all the economic forecasts agree on the heavy consequences the pandemic is going to have - not only in the short run - on global GDP, consumption, employment and stock market values, even if quantitative estimates of the impact and its distribution over time are still quite

uncertain[1] (Chudik et al., 2020, Baldwin and Weder di Mauro, 2020). The studies
emphasize similarities and, above all, differences of a global pandemic, compared to previous infectious diseases (Spanish Flu, Ebola virus, Sars, HIV/Aids) and try to estimate the potential economic losses caused by COVID-19 under different scenarios. In fact, when the simulations are carried out, researchers can only rely on data from the first infection wave and the content and duration of restriction measures are still largely unpredictable (Deb et al., 2020): that is why, in some cases, the GDP growth in 2020 is expected between -4.5% and -12.9% in Italy and between -3.1% and -9.7% in Japan, to be compared to a median GDP loss of the 30 major economies between 2.8% and 10.7% (Fernandes, 2020); while other studies predict GDP losses in 2020 between 6 and 214 billion dollars in Italy and between 17 and 549 billion dollars in Japan (McKibbin and Fernando, 2020).

Despite the evidence that all the sectors are involved in the economic slowdown, the analyses agree on expecting heavier losses in the service sector (travel, tourism and hospitality related activities) and in international trade (Nicola et al., 2020).

5.2   The Effect of Lockdown on Gross National Happiness Index

Greyling, Rossouw, and Adhikari (2020) studied the impact on the GNHI of Section 3.3 of the lockdown measures. The authors used daily data for the time period from 1 January to 8 May 2020. Figure 5.1 shows a negative peak on the date of the lockdown but, from that moment on, then, the index reverts to its average, something quite different from what we will observe in Section 5.5 in which a persistent reduction of the well-being indexes for both Italy and Japan is assessed.

The primary goal of the study was to determine the effect of the pandemic on happiness before and after the implementation of regulations,aiming at curbing the spread of the virus. They focus their attention on South Africa and therefore they introduce a dummy variable `lockdown` which is set equal to 1 after 18 March and 0 before that date. As control covariates they included the `consumption` as a proxy for private current expenditure and, in turn, consumption is estimated through the daily data available on credit and debit card sales together with ATM transactions (BETI , 2020). As no daily measure of unemployment is available, the authors applied the methodology as set out by Nuti et al. (2014) and Brodeur et al. (2020) and use daily searches

---

[1]While we write these notes, the second wave of the virus spread is still ongoing and its aftermath will depend, among other things, on the efficacy of the anti-crisis measures.

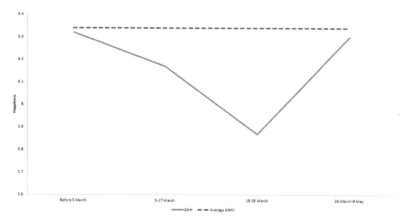

Figure 5.1
Happiness level before and after the lockdown of 18 March 2020 as seen from the GNH index. Source: Adapted from Figure 3 from Greyling, Rossouw, and Adhikari (2020).

on Google Trends for `job` as a proxy for future job uncertainty. To proxy the sale of `alcohol` and `tobacco` they apply a similar strategy used for the `job` variable, finally looking for the search of the word "beer" as a proxy for both.[2] They also include concerns about the schooling of children and the lack of mobility using respectively Google Trends and the Google Mobility Reports.[3] The authors also noticed an increase in the volume of tweets during the lockdown as a proxy for connectivity. They finally include the number of Covid cases and its square in the model. They estimate two different models. The first one is a standard OLS model for y_t as the GNH Index for day t:

$$y_t = \alpha_0 + \alpha_1 \texttt{lockdown}_t + \alpha_2 X_t + \epsilon_t \tag{5.1}$$

where X_t is the set of control covariates described in the above. The second one is a glm model of this form:

$$P(\texttt{Happy} = 1 | X_t) = \alpha + \beta \texttt{lockdown}_t + \beta_1 X_t + \epsilon_t \tag{5.2}$$

where `Happy` is a dummy variable that takes value 0 if GNH index is lower than its average value (6.35) in the year 2019. The second model tries to estimate the probability of being happier than the average level of 2019 before and after the lockdown.

Table 5.1 summarizes the results of model (5.1). The statistically significant determinants, which are negatively related to happiness, show the expected relationships. In particular, *i)* if searches for jobs increases happiness

---

[2] Apparently in South Africa, "beer" is a good proxy for "alcohol" and Google searches for "tobacco" are correlated with those of "beer" with correlation equal to 0.83.
[3] Google LLC "Google COVID-19 Community Mobility Reports", https://www.google.com/covid19/mobility/

decreases; *ii)* when the number of tweets increases, happiness decreases; *iii)* more searches for "beer" implies less happiness; *iv)* searches for school are related to lower levels of happiness and *v)* the daily number of COVID-19 cases impacts negatively the happiness level.

As for the model in (5.2), the same evidence is confirmed in terms of the coefficients. What the authors found was that the probability of being happy, i.e., $P(\text{Happy} = 1|X_t)$, is 0.23 for the full sample, 0.26 before the lockdown and 0.17 after the lockdown. Moreover, by simulating a "no lockdown" scenario, this probability goes up to 0.27, all estimates being statistically significant.

Table 5.1
Results for the estimation of model (5.1) adapted from Table 2 in Greyling, Rossouw, and Adhikari (2020). *Full* refers to the whole data, *Before* and *After* refers to the data before and after the lockdown. Standard errors of estimates in parentheses.

Variable	Full	Before	After
Lockdown	−0.2656**		
	(0.1066)		
Log(sales)	0.1264**	0.2115***	0.0300
	(0.0499)	(0.0512)	(0.0679)
Job	−0.0009*	−0.0001	−0.0004*
	(0.0014)	(0.0015)	(0.0004)
Log(tweets volume)	−0.3633*	−0.1532	−0.7012**
	(0.1956)	(0.3505)	(0.3338)
Alcohol	-0.0104***	−0.0085	−0.0082***
	(0.0017)	(0.0060)	(0.0014)
School	−0.0031***	0.0432	−0.0030**
	(0.0010)	(0.0800)	(0.0012)
COVID-19 cases	−0.0015**	−0.0300**	−0.0010*
	(0.0007)	(0.0141)	(0.0009)
COVID-19 cases squared	0.00001****	0.0000	0.0000*
	(0.0000)	(0.0005)	(0.0000)
Retail	−0.0048**		
	(0.0019)		
Constant	9.5322***	6.9730**	9.7575***
	(2.2862)	(3.1482)	(2.8918)
N	128	77	51
Adjusted R^2	0.368	0.330	0.667

Note: $^*p<0.1$; $^{**}p<0.05$; $^{***}p<0.01$

5.3   Hedonometer and the COVID-19 Pandemic

The Hedonometer project found that, although the COVID-19 caused a large dip in happiness, over time happiness has rebounded to be close to levels before COVID-19, asides from the decrease in happiness caused by the murder of George Floyd and the consequent protests against police brutality. At least for the case of all English tweets. The situation is similar for the French, Spanish ad German tweets with the difference that the lowest peak are reached at the beginning of the pandemic around March 2020 and then slowly the indexes recover to pre-pandemic levels (see Figure 5.2). It seems, as noticed in Section 3.2 that the instrument is too sensitive to the reference dictionary used to classify the data to be able to capture long or medium term effects of the pandemic on well-being.

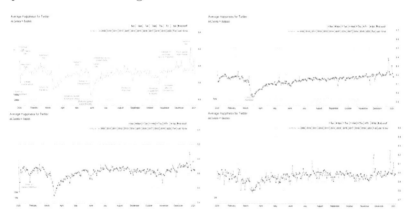

Figure 5.2
Happiness level in 2020 according to the Hedonometer for tweets written in English, Spanish, French and German. Source: `https://hedonometer.org/`

5.4   The World Well-Being Project and Tracking of Symptoms During the Pandemic

Guntuku et al. (2020) make use of the World Well-Being Project data to track mental health and symptom mentions on Twitter during the COVID-19 pandemic. As an effect of the magnitude of the novel coronavirus (COVID-19) pandemic, considerable economic hardships, stress, anxiety, and concerns about the future have been measured by these authors through Twitter. The authors tracked the change of language related to the description of mental and physical health by Twitter users in the United States from January to May 2020. Using the approach presented in Section 3.4, they extract the relative frequency of single words and phrases and, based on the word and phrase frequencies related to mental health, they estimate the corresponding level of sentiment, stress, anxiety and loneliness. These estimates are tracked over time to spot any impact of the national declaration of emergency, on March 13, to May 6, and compare them to the estimates from the same period in 2019, controlling for day of the week and seasonality effects. Figure 5.3 shows the results visually. In summary, mental health estimates in the duration after the declaration of emergency from March 13 to May 6, show that: sentiment was lower in 2020 compared with that in 2019, stress was higher, anxiety was consistently higher, and loneliness also showed a marked increase. Symptom mentions in the COVID-19 related tweets captured also emerging symptoms such as a change in smell/taste, body aches and skin lesions. The results are shown in Figure 5.4.

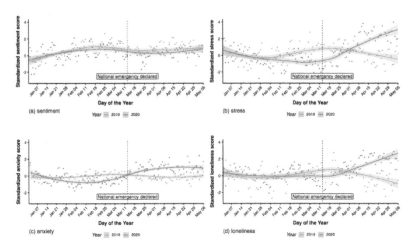

Figure 5.3
(a) sentiment, (b) stress, (c) anxiety and (d) loneliness expressions derived from data-driven machine learning models on Twitter language from the start of January till May 6 in 2019 (green) and 2020 (orange). The measures are normalized by centering and scaling based on January values of the respective years and calculating the mean over all states in the USA weighted by the number of Tweets in each state. Source: Figure 1 in Guntuku et al. (2020).

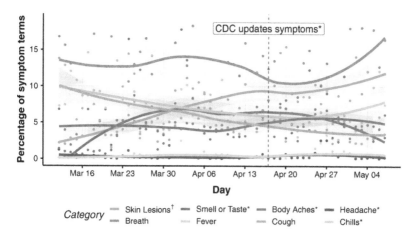

Figure 5.4
Trends in symptom mentions in COVID-19 related tweets. *Smell/taste, body ache, headache, chills were added to the symptom list by the Centers for Disease Control (CDC) on April 17. †Skin lesions are increasingly being discussed in the context of COVID-19 tweets. Source: Figure 2 in Guntuku et al. (2020).

5.5 The Decline of SWB-I & SWB-J During COVID-19

As in Section 3.5.6, the data used in this chapter come from two different repositories that were collected under two different projects but in both cases using Twitter search API. The Japanese tweets were collected using only the filter on language = **Japanese** and country = **Japan** and similarly for Italy (**Italian** and **Italy**). According to Statista[4] there are about 8 million accounts active daily in Italy whilst about 52 millions are in Japan, therefore the number of tweets posted is not comparable. To keep the volumes of tweets comparable between the two indicators we imposed a maximum number of at most 50,000 tweets per day to our Twitter crawlers. As a result, the total volume of tweets is 13,975,242 for Italy and 12,907,902 for Japan and the data were collected since 2020-11-01 till 2020-10-11 for Italy and 2020-09-20 for Japan. These tweets are part of two separate repositories that were collected for Italy since 2012 and for Japan since 2015. For both projects, systematic download of data was stopped in 2018[5] and resumed on late 2019 with alternate fortune due to changes in Twitter API limits. For Italy some historical data were collected ex-post for the year 2019 for other research repositories and included in this data collection. Table 5.2 shows the yearly average values

---

[4]https://statista.com
[5]These represent the collection of data of the previous two chapters.

Table 5.2
Average values of SWB-I and SWB-J from 2015 till 2020. For Italy, data in 2015 were not available for the whole year. The statistics for 2019 are referred only to the months of November and December. For 2020, the average refers to period from 1st January up to 11th October for Italy and 20th September for Japan.

Year	2012	2013	2014	2015	2016	2017	2018	2019	2020	2019 vs. 2020
SWB-I	48.9	52.2	49.7	48.7	50.5	57.7	55.7	54.1	42.4	−11.7
SWB-J	–	–	–	54.4	53.6	53.2	52.5	35.3	27.0	−8.3

of the SWB-I and SWB-J indicators since 2012 and 2015 for Italy and Japan respectively.

In both cases, the two indicators dramatically dropped (more in Italy than in Japan) in 2020. While the average values can be attributed intuitively to cultural differences in the way positive and negative emotions are expressed in the two countries, clearly the 2020 drop is definitely related to the COVID-19 pandemic, yet with differences between the two countries as it will be explained in the next sections.

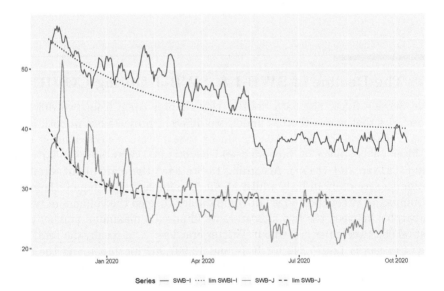

Series —— SWB-I ···· lim SWBI-I —— SWB-J – – lim SWB-J

Figure 5.5
SWB-I and SWB-J indexes from November 2019 till 10 October 2020 for Italy and 20th September for Japan with estimated limiting dynamical systems.

In order to compare the two time series in a stochastic modelling approach, we try to fit four different stochastic differential equations models (Iacus, 2008,

Iacus and Yoshida, 2018). As the two time series exhibit a negative trend in the first half of 2020 and seem to rebound around a long term mean, we assume a drift coefficient of mean-reverting time. A typical drift form is described as $\alpha(\beta - X_t)$ where β represents the long term mean around which the time series X_t oscillates and α is called the *speed* of mean reversion: the higher α, the faster the process converges to its long run mean. A first example of this model is the so called Ornstein-Uhlenbeck (Uhlenbeck and Ornstein, 1930) or Vašíček model (Vasicek, 1977), which is a continuous version of an AR(1) model:[6]

$$dX_t = \alpha(\beta - X_t)dt + \sigma dW_t, \quad X_0 = x_0 \tag{5.3}$$

where dX_t is infinitesimal increment of the process between t and $t + \delta$, i.e., $dX_t = X_{t+\delta} - X_t$, $dW_t \sim N(0, dt)$ is the infinitesimal increment of the Wiener process (or Brownian motion) that represents exogenous random Gaussian shocks, $\sigma > 0$ is the scaling factor and x_0 is some initial condition. Next is the geometric Brownian motion (Black and Scholes, 1973, Merton, 1973) model:

$$dX_t = \alpha(\beta - X_t)dt + \sigma X_t dW_t, \quad X_0 = x_0 \tag{5.4}$$

where the term $\sigma X_t dW_t$ represents the feedback in the system, in the sense that the noise dW_t interacts with the state of the system X_t. We also fit the CIR (Cox-Ingersol-Ross) model (Cox, Ingersoll, and Ross, 1985) which differs from the previous model for the $\sqrt{X_t}$ in the diffusion coefficient.[7] The role of the square root is to dump a bit the feedback effect:

$$dX_t = \alpha(\beta - X_t)dt + \sigma\sqrt{X_t}dW_t, \quad X_0 = x_0. \tag{5.5}$$

And finally, when one is uncertain about the dumping effect of the feedback term, one can generalize the model to the CKLS (Chan, Karolyi, Longstaff and Sanders) model (Chan et al., 1992) adding an exponent $0 < \gamma < 2$ in the diffusion coefficient:

$$dX_t = \alpha(\beta - X_t)dt + \sigma X_t^\gamma dW_t, \quad X_0 = x_0 \tag{5.6}$$

Clearly, CKLS embeds VAS (for $\gamma = 0$), CIR (for $\gamma = 0.5$) and GBM (for $\gamma = 1$). We estimate these models on the two SWB-I and SWB-J indexes thorough the `yuima` R package (Brouste et al., 2014, Iacus and Yoshida, 2018) via quasi-maximum likelihood estimation. The results of the estimation are given in Table 5.3. The main characteristics of the first threee models is that VAS model has Gaussian increments, GBM has log-normal increments and CIR has non-central chi-squared increments.

Looking at the results in Table 5.3 it seems clear that the models agree on the α and β parameters. Indeed, β correspond to about 39 for SWB-I and 28 for SWB-J across models with about 11 percentages points of difference

---

[6]For $\beta = 0$.

[7]In a stochastic differential equation model the term associated to dW_t is called *diffusion* coefficient.

Table 5.3
Fitting different SDE models. Standard errors of the estimates are in parentheses and (·) means estimates of variance-covariance matrix did not converge).

Index	α	β	σ	γ	Model	AIC
	3.16	38.99	14.7		VAS	787.0
	(2.41)	(6.22)	(0.56)			
	3.57	39.54	0.33		GBM	782.9
	(2.46)	(4.60)	(0.01)			
SWB-I						
	3.34	39.28	2.20		CIR	781.9
	(2.42)	(5.35)	(0.08)			
	3.42	39.37	1.12	0.68	CKLS	783.6
	(2.44)	(5.08)	(·)	(·)		
	12.92	28.44	26.25		VAS	1,090.0
	(5.52)	(2.19)	(1.05)			
	11.46	28.43	0.83		GBM	1,027.4
	(5.71)	(2.23)	(0.03)			
SWB-J						
	11.98	28.42	4.62		CIR	1,055.1
	(5.61)	(2.21)	(0.19)			
	11.64	28.41	0.05	1.84	CKLS	1,010.7
	(5.93)	(2.22)	(0.01)	(0.08)		

between the two countries, and α varies in the interval [3.16, 3.57] for SWB-I and in the interval [11.46, 12.92] for SWB-J, meaning that the convergence to the long run mean of the SWB index is between three and four times faster in Japan than in Italy. Figure 5.5 show also the estimated limit deterministic dynamical systems obtained taking the limit as $\sigma \to 0$ of the stochastic differential equation models which best fits, in terms of AIC, the set of data (respectively, CIR for SWB-I and CKLS for SWB-J): what these deterministic limit processes show is that the decline of subjective well-being in Italy is much more emphasized, in 2020, compared to what happens in Japan whilst the convergence to the long run mean is faster for Japan.

It is worth to remark that these results are more in line with Section 5.4 than with the evidence from Hedonometer analysis of Section 5.3 and the Gross National Happiness Index of Section 5.2.

In summary, the analysis shows that the subjective well-being has a short run (or volatile) and a medium-long run (or structural) component, which are usually intended as emotional well-being *versus* life evaluation. Moreover, the mean is not necessarily stable during the whole year and therefore, a annual single survey and the consequent point estimate are not appropriate to capture the overall level and evolution of well-being of a country. All the indexes presented capture one or both characteristics.

5.5.1 Related Studies

Although not based on Twitter data, but strictly related to the topic of this chapter is a survey measure of subjective well-being which, along with Google Trends data, underpins the work by Foa, Gilbert, and Fabian (2020), where the authors examine the evolution of well-being in Great Britain during the first wave of the pandemic, in spring 2020. The original contribution of this study is in his attempt to disentangle the well-being effect of the pandemic from the effect of the lockdown restrictions, showing that, while the COVID-19 outbreak negatively affects self-perceived well being, the imposition of lockdown measures has a positive impact on well-being, likely due to a better work-life balance, an increase in remote worker autonomy, the adoption of government support schemes. Google Trends series allow for an extension of the results to countries other than Great Britain.

The significant shift in households' economic sentiment in the 27 EU countries, following the COVID-19 outbreak and ensuing lockdown restrictions, is investigated by van der Wielen and Barrios (2020). The research, similarly to the previous ones, relies on panel dataset coming from Google Trends, covering business cycle, labor market and consumption related queries. As a result, the study reveals a substantial worsening of the sentiment along all the investigated dimensions in the months after the pandemic outbreak. The change of the sentiment is more pronounced in the EU countries where the economic and labor market conditions were less favorable at the onset of the health crisis, arguably reflecting the fear of a persistent high unemployment level in the post-pandemic phase. Finally, in a comparison with the consequences of the 2008 economic recession, the study shows that the fall in economic sentiment is even more marked during the 2020 pandemic, especially for unemployment-related sentiment, confirming widespread concerns about pervasive and long-lasting consequences of the crisis.

5.6 Data Collection of Potential Determinants of the SBW Indexes

Apart from noting a decline of the SWB indexes for both Italy and Japan, this study aims at understanding the impact of several potentially explanatory factors on the SWB indicators for these two countries. For this reason, several data sources have been considered.

5.6.1   COVID-19 Spread Data

On the official statistics side, we obtained from the WHO the data on the number of confirmed COVID-19 cases and deaths. We replaced negative values[8] with zeros and take 7-days moving average to reduce the impact of weekend days late reporting, that induces artificial periodicity in the data.

5.6.2   Financial Data

In order to capture high frequency financial dynamics, we consider the main stock market indexes of both countries, specifically the Nikkei and FTSE MIB index. Daily adjusted closings data are taken from Yahoo! Finance[9] through the `quantmod` package (Ryan and Ulrich, 2020).

5.6.3   Air Quality Data

We then considered air quality data available through the Air Quality Open Data Platform[10] and, in particular, the PM2.5 pollutant concentration and the temperature. Also in this case we considered a 7-days moving average and aggregate data at country level. The two variables roughly capture, on one hand, the amount of pollutant reduction during the pandemic due to the lockdown and, on the other hand, the effect of high/low temperature on mood (Curini, Iacus, and Canova, 2015).

5.6.4   Google Search Data

Subjective well-being is influenced by micro and macro variables, where the latter are usually captured by official statistics. Both micro and macro observable dimensions are characterized by very low frequency and are available with considerable delay. Therefore, as in Choi and Varian (2012) we adopt as proxies of these variables several types of Google research topics in the two countries available through the Google Trends[11] portal. Data have been downloaded through the `gtrendsR` package[12] (Massicotte and Eddelbuettel, 2020).

Google Trends offers two types of search statistics: one is based on the exact keyword and one is based on the concept of *topic*. The difference is that topics include all search terms related to that topic or, put it another way, topics are a collection of search terms. Topics are normalized across countries,

---

[8]Negative values exists because the number of cases is obtained through differentiation of the cumulative number of cases, that is updated retrospectively in some cases, leading to negative differences.

[9]https://finance.yahoo.com

[10]World Air Quality Index (WAQI) https://aqicn.org/data-platform/covid19/

[11]Google Trends, https://www.google.com/trends

[12]All the computations have been processed using the R statistical environment (R Core Team, 2020).

so there is no need to translate a topic into a specific language. We included several research topics, distinguishing between general web searches (blogs, forum, etc.) and specific news-related searches. We included terms related to the pandemic, real economy and job market, health conditions and searches for adult content. The latter has been introduced following the suggestions in by Stephens-Davidowitz (2018) that found that Google search for adult content predicts the United States unemployment rate from 2004 into 2011 with the motivation that *"(...) unemployed people presumably have a lot of time on their hands. Many are stuck at home, alone and bored"*. Indeed, *"with the global expansion of the COVID-19 pandemic, social or physical distancing, quarantines, and lockdowns have become more prevalent and concurrently, Pornhub, one of the largest pornography sites, has reported increased pornography use in multiple countries, with global traffic increasing over 11% from late February to March 17, 2020"* (Mestre-Bach, Blycker, and Potenza, 2020). Happyness literature also seem to consider this relationship with, so to say, boring living conditions (D'Orlando, 2011) as well as to COVID-19 specific stress conditions (Döring, 2020).

Table 5.4 contains the complete list of topics. Topics and keywords are the same for the two countries with two differences: 1) for Japan only we added the search for the keyword Corona "コロナ" in katakana alphabet, as we noticed a remarkable difference between the topic and this exact search in terms of time series patterns; 2) for Italy only we included the keyword "Rt" for reproduction number, as this is often reported in the news and in the official statements of the government. The Rt indicator is not associated to the virus in Japan, so we did not include it for Japan.

Table 5.4

Google search topics used in the analysis.

Search term	Country	Keyword type	framework	Type of search
Coronavirus	IT/JP	topic	pandemic	web
CoronavirusNews	IT/JP	topic	pandemic	news
(コロナ) Corona	JP	topic	pandemic	web
(コロナ) CoronaNews	JP	topic	pandemic	news
Covid	IT/JP	topic	pandemic	web
CovidNews	IT/JP	topic	pandemic	news
Rt	IT	keyword	pandemic	web
Wuhan	IT/JP	topic	pandemic	web
Unemployment	IT/JP	topic	economics	web
UnemploymentNews	IT/JP	topic	economics	news
Economy	IT/JP	topic	economics	web
EconomyNews	IT/JP	topic	economics	news
GDP	IT/JP	topic	economics	web
GDPNews	IT/JP	topic	economics	news
Depression	IT/JP	topic	health	web
Stress	IT/JP	topic	health	web
Insomnia (disorder)	IT/JP	topic	health	web
Health	IT/JP	topic	health	web
Solitude	IT/JP	topic	health	web
AdultContent	IT/JP	topic	leisure	web

5.6.5 Google Mobility Data

In addition to the Google search data, we include the human mobility data obtained through the Google COVID-19 Community Mobility Reports.[13] We considered a 7-days moving average data of the "residential and workplace percent change from baseline" statistic available in the data to capture, roughly, the effect of lockdown restrictions on human mobility.

5.6.6 Facebook Survey Data

Since late April 2020 Facebook, in partnership with several universities, has conducted the "COVID-19 World Survey Data" collecting several indicators. These COVID-19 indicators are derived from global symptom surveys that are placed by Facebook on its platform. The surveys ask respondents how many people in their household are experiencing COVID-like symptoms, among other questions. These surveys are voluntary, and individual survey responses are held by University of Maryland and are shareable with other health researchers under a data use agreement. No individual survey responses are shared back to Facebook. Using this survey response data, the University of Maryland estimated the percentage of people in a given geographic region on a given day who:

- have COVID-like illness (FB.CLI) = fever, along with cough, shortness of breath, or difficulty breathing;

- have influenza-like illness (FB.ILI) = fever, along with cough or sore throat;

- have reported to use mask cover (FB.MC);

- have reported had direct contact (FB.DC), longer than one minute, with people not staying with them in last 24 hours;

- are worried about themselves and their household's finances in the next month (FB.FH). Only respondents who have replied as being very worried and somewhat worried.

Instead of direct counts that may have missing data for some date, we use the smoothed versions of the indicators based on a seven-day rolling average. Data have been collected through the COVID-19 World Symptom Survey Data API (Fan, Li, et al., 2020).

5.6.7 Restriction Measures Data

Finally, we constructed a dummy variable `lockdown` for both countries, taking value 1 when lockdown or other types of restrictions were in force in each

---

[13]Google LLC "Google COVID-19 Community Mobility Reports", `https://www.google.com/covid19/mobility/` Accessed: 2020-10-15.

country. For Italy,[14] a national lockdown was enforced since 9 March 2020 and lifted on 3 June 2020. For Japan,[15] there was no strict lockdown, but the state of emergency has been declared starting from 8 April 2020 and lifted on 21 May 2020 for most prefectures. But, as the remaining five prefectures has to wait till 25 May 2020, we decided to set our dummy equal to one for Japan for the whole period 2020-04-08/2020-05-25.

Table 5.5 reports the complete list of variables used in the analysis. It is worth mentioning that, due to the different time coverage of the data, we extended the analysis to 10 October 2020 for Italy and to 20 September 2020 for Japan.

Table 5.5
Times series used in the analysis. In the analysis Japanese variable start with *j* and the Italian ones with *i*, e.g., iCases, jCases.

Variable	Area	Source
SWB-I, SWB-J	subjective well-being	Twitter
Solitude	well-being	Google Trends
Depression	well-being	Google Trends
Stress	well-being	Google Trends
Insomnia	well-being	Google Trends
Health	health/well-being	Google Trends
PM2.5	health/environment	WAQI
Temperature	environment	WAQI
Cases	pandemic	WHO
Deaths	pandemic	WHO
Coronavirus, CoronavirusNews	pandemic	Google Trends
(コロナ) Corona, CoronaNews	pandemic	Google Trends
Covid, CovidNews	pandemic	Google Trends
Rt	pandemic	Google Trends
Wuhan	pandemic	Google Trends
Unemployment, UnemploymentNews	economy	Google Trends
Economy, EconomyNews	economy	Google Trends
GDP, GDPNews	economy	Google Trends
FTSEMIB	economy	Yahoo! Finance
Nikkei	economy	Yahoo! Finance
AdultContent	leisure	Google Trends
FB.CLI	health/well-being	Facebook
FB.ILI	health/well-being	Facebook
FB.MC	behavioral	Facebook
FB.DC	behavioral	Facebook
FB.FH	well-being	Facebook

5.7 What Impacted the Subjective Well-Being Indexes?

Although the SWB-I and SWB-J indicators are composed by a range of different dimensions, it is also true that components like **emo** may be affected

---

[14]Source: https://en.wikipedia.org/wiki/COVID-19_pandemic_in_Italy
[15]Source: https://en.wikipedia.org/wiki/COVID-19_pandemic_in_Japan

by the pandemic in several ways, like, e.g., stress, fear of the virus coming from the news or the statistics of cases and deaths, etc. In this study, we want to investigate and possibly describe the complexity of the determinants of subjective well-being, as measured by the proposed social media indexes, following a data science approach.

5.7.1 Preliminary Correlation Analysis

A simple correlation analysis[16] between each index and the other covariates may lead to counter-intuitive conclusions. Indeed, looking at Figure 5.6 it is possible to notice that for some variables the correlation with SWB-I (and SWB-J) change monthly, from November 2019 till September 2020, and also with respect to the whole 2020 (last row of each correlation plot in Figure 5.6). More precisely, looking at Figure 5.6 top panel for SWB-I, while the variable temperature (`itemperature`) is negatively correlated with well-being in Italy, monthly and for the whole year 2020, news about economy (`iEconomyNews`) have positive correlation up to May 2020, and then it switches sign. Or, again, the variables `iCoronaVirus`, `iCoronaVirusNews`, `iCovid`, `iCovidNews`, `iHealth` are negatively correlated between January and March 2020, then almost all the correlations switch sign. Other variables, like `ilockdown` do matters, with negative correlation, but clearly only in the periods when the lockdown is actually in force. Similar up-and-down trends occur for the Japanese index. It is quite surprising, when we read the panels of Figure 5.6 by row, noting that groups of variables that show a positive relationship with SWB at some point in time change the correlation sign in a month or so. Clearly, this is the effect of the complexity of the well-being itself and its dynamics over time. More details emerge from the monthly regression analysis.

5.7.2 Monthly Regression Analysis

To take into account the dynamic nature of subjective well-being, shown also by the preliminary correlation study, we run a monthly stepwise regression analysis.[17] Tables 5.6-5.7 report the estimated coefficients and summary analysis for the SWB-I index, whilst Tables 5.8-5.9 show the same analysis for the SWB-J index. In order to save space, the tables show only the months from January to September 2020. The tables also show the results for the whole period January–September 2020.

　　Considering the Italian case first, it can be seen that the following variables have a negative effect on the yearly scale: `iRt`, `iGDP`, `iHealth`, `iInsomnia`, `iWuhan`, `ilockdown`, `iFB.CLI` and `iFB.DC` have a negative impact on SWB-I,

---

[16]For the analysis we used the Spearman correlation coefficient that is a rank-based measure of association. This measure is known to be more robust and it possibly captures also association between non-linear, yet monotonic, transformation of the data.

[17]As the data set contain 27 covariates and only 30 days of observations, we were forced to run a step-wise regression analysis.

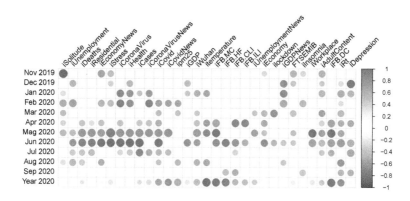

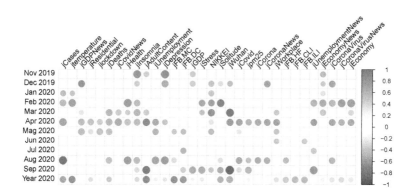

Figure 5.6
Correlation analysis of SWB-I (top panel) and SWB-J (bottom panel) versus all the covariates used in the analysis, by month and for the year 2020 (January to September). Spearman correlation coefficient, only shown statistically significant estimates with p-values <0.05.

Table 5.6

Monthly step-wise regression analysis for SWB-I for the months January, February, March and April and for the full period January-September 2020.

	Dependent variable: SWB-I				
	Jan–Sep	Jan	Feb	Mar	Apr
iDeaths			−98.06**		0.62*
			(39.87)		(0.32)
iCases			15.76		−1.35***
			(10.39)		(0.35)
FTSEMIB		−0.39*	0.68***	0.31***	0.23
		(0.23)	(0.23)	(0.10)	(0.20)
iCoronaVirus			5.11*	−3.81***	162.18**
			(2.90)	(0.62)	(55.84)
iCoronaVirusNews	0.19***	−0.84**	−13.02*	−2.03	−146.85**
	(0.05)	(0.32)	(6.69)	(1.24)	(51.55)
iCovid	0.09***		47.16*	41.20***	−93.24**
	(0.03)		(25.16)	(9.94)	(35.02)
iRt	−0.05***	−0.96**	−12.90		8.29**
	(0.01)	(0.41)	(9.99)		(3.13)
iCovidNews				−68.92***	−59.26***
				(17.28)	(19.71)
iUnemployment	0.05***	−0.41**	−19.96*	1.22***	0.99***
	(0.02)	(0.18)	(11.06)	(0.28)	(0.33)
iGDP	−0.11***				
	(0.03)				
iGDPNews	0.03***				
	(0.01)				
iHealth	−0.19***				
	(0.05)				
iInsomnia	−0.02				
	(0.02)				
iResidential	0.18***		3.73		
	(0.05)		(2.80)		
iWorkplace	0.16***		5.27	−0.57	
	(0.06)		(4.29)	(0.36)	
ipm25				−0.57	−0.80*
				(0.36)	(0.40)
itemperature	0.11***		0.70**	0.98	2.41**
	(0.04)		(0.28)	(0.62)	(0.99)
iWuhan	−0.05***				
	(0.01)				
ilockdown	−0.07***			−0.11*	
	(0.02)			(0.05)	
iFB.CLI	−0.06***				−0.84**
	(0.02)				(0.34)
iFB.ILI	0.04*				1.03**
	(0.02)				(0.35)
iFB.MC					9.49***
					(2.96)
iFB.DC	−0.27***				−64.48**
	(0.05)				(23.61)
iUnemploymentNews		0.21**			
		(0.09)			
iFB.HF					
swbiLag	0.82***	0.71***		0.30*	0.38*
	(0.03)	(0.14)		(0.17)	(0.20)
Observations	285	31	29	31	30
R^2	0.99	0.99	1.00	0.99	0.99
Adjusted R^2	0.99	0.98	0.99	0.98	0.97

Note: $^*p<0.1$; $^{**}p<0.05$; $^{***}p<0.01$

Table 5.7

Monthly step-wise regression analysis for SWB-I for the months May, June, July, August and September 2020.

	May	Jun	Jul	Aug	Sep
			SWB-I		
	May	Jun	Jul	Aug	Sep
iDeaths	−1.67***		−6.57***		
	(0.25)		(2.13)		
iCases	1.34***	−2.44***		1.91***	−2.39*
	(0.39)	(0.61)		(0.65)	(1.30)
FTSEMIB	0.19	−0.20***		0.20	
	(0.16)	(0.06)		(0.17)	
iCoronaVirus	13.58***	23.87***	8.88***	−49.04**	−24.12***
	(2.15)	(2.99)	(2.01)	(18.71)	(7.31)
iCoronaVirusNews	8.05**	−12.51***	−3.15***	50.25***	37.32***
	(2.91)	(2.25)	(0.40)	(17.05)	(7.07)
iCovid	−3.18***			−4.91**	2.87***
	(0.89)			(2.26)	(0.92)
iRt	0.13**	0.73		1.00*	3.23***
	(0.06)	(0.46)		(0.49)	(0.82)
iCovidNews		−6.22***		5.55**	−5.30***
		(1.11)		(2.16)	(1.00)
iUnemployment	1.24*	−1.47**			−8.90***
	(0.70)	(0.58)			(1.97)
iGDP					
iGDPNews					
iHealth					
iInsomnia					
iResidential	−0.90***	−1.70***		−2.11***	4.17**
	(0.31)	(0.55)		(0.69)	(1.50)
iWorkplace				0.39*	−3.54**
				(0.22)	(1.41)
ipm25	−0.56**	−0.26		1.42**	
	(0.20)	(0.25)		(0.58)	
itemperature		−1.04***	−1.46***	−3.74***	2.21**
		(0.34)	(0.27)	(1.14)	(1.00)
iWuhan					
ilockdown		0.08***			
		(0.02)			
iFB.CLI	0.04				0.42*
	(0.03)				(0.20)
iFB.ILI			0.13***		−0.45**
			(0.03)		(0.18)
iFB.MC	2.43***	0.25	0.76***		
	(0.34)	(0.23)	(0.23)		
iFB.DC		1.68**		1.78	5.27**
		(0.75)		(1.25)	(1.98)
iUnemploymentNews				1.23***	
				(0.39)	
iFB.HF	−0.03	−0.31**		0.63**	
	(0.02)	(0.12)		(0.26)	
swbiLag	−0.40***	−0.27*		−0.25	−0.30
	(0.13)	(0.14)		(0.15)	(0.19)
Observations	31	30	31	31	30
R^2	1.00	1.00	1.00	1.00	1.00
Adjusted R^2	1.00	1.00	1.00	1.00	0.99

Note: *$p<0.1$; **$p<0.05$; ***$p<0.01$

Table 5.8

Monthly step-wise regression analysis for SWB-J for the months January, February, March and April and for the full period January–September 2020.

| | Dependent variable: | | | | |
| | SWB-J | | | | |
	Jan–Sep	Jan	Feb	Mar	Apr
jDeaths	0.07** (0.03)	3.16** (1.15)	6.22* (3.43)	1.01 (0.76)	0.12 (0.07)
jCases	−0.15*** (0.04)		−23.28** (9.76)		−1.21 (0.83)
NIKKEI					−0.40** (0.15)
jCoronaVirus	−0.16*** (0.05)			1.54** (0.59)	(0.68)
jCoronaVirusNews	−0.07*** (0.03)	0.91** (0.33)	2.75** (1.09)	2.71*** (0.76)	−10.60** (3.93)
jCorona	0.17*** (0.05)		12.04 (7.69)	6.11*** (2.12)	3.41*** (1.02)
jCovidNews	0.04 (0.03)		1.78 (1.50)	1.57*** (0.40)	
jUnemployment	−0.12** (0.06)				
jUnemploymentNews	0.06*** (0.02)				
jEconomy	0.09** (0.05)				
jGDPNews	−0.03* (0.02)				
jStress	−0.07*** (0.02)				
jDepression	0.13*** (0.04)				
jHealth	−0.07* (0.04)				
jSolitude	−0.08*** (0.02)				
jInsomnia	−0.09** (0.04)				
jResidential	−0.20** (0.10)		0.92 (0.61)		
jWorkplace	−0.11 (0.07)				−0.87** (0.32)
jtemperature	0.18* (0.09)		1.26*** (0.42)		−1.12** (0.43)
jlockdown					−0.09* (0.05)
jFB.CLI	0.20** (0.10)				−0.53** (0.20)
jFB.ILI	−0.20*** (0.08)				0.67** (0.23)
jFB.MC					−1.00* (0.49)
jFB.DC	−0.15** (0.06)				0.81*** (0.22)
jFB.HF	0.10*** (0.03)				
jpm25				0.31*** (0.10)	
jCoronaNews		−2.94*** (1.00)	−14.29* (7.91)	−8.53*** (3.01)	−10.24*** (3.15)
jCovid			11.59** (4.94)	−9.88*** (2.74)	31.64** (14.47)
swbjLag	0.77*** (0.03)	0.72*** (0.11)	0.64*** (0.21)		0.43** (0.20)
Observations	264	31	29	31	30
R^2	0.94	0.87	0.96	0.92	0.98
Adjusted R^2	0.93	0.86	0.94	0.89	0.96

Note: $^*p<0.1$; $^{**}p<0.05$; $^{***}p<0.01$

which is quite admissible, but iCoronavirusNews, iCovid, iUnemployment, iGDPNews, iResidential, iWorkplace, itemperature, iFB.ILI have a positive impact on SBW-I and this is probably the outcome of an average effect, though we should remind that the coefficients are standardized, hence they are comparable to each other in magnitude. In this respect, the negative coefficients are on average larger than the positive ones, which may ease the comprehension of this output. In all events, a deeper look at time evolution of the pandemic can inform better on the role of these factors in explaining the variance of the SWB-I. In fact, if we look at the variable iDeaths, we may note that its coefficient is strongly negative in February and also negative in May and July while iCases has a negative impact in April and June with a coefficient higher, in absolute terms, than iDeaths for the same months (when the effect of iDeaths is positive): this indicates that the reactions to the COVID-19 outbreak change over time, following the relative intensity of the phenomena (affections, casualties) and are, to some extent, substitutes. On the other hand, the iCovid coefficient is negative in April, May, June and August and iCoronaVirusNews is negative from January till April and then again in July and so forth. This means that the news about the pandemic (iCovid, iCoronaVirusNews, etc) and the effect of the pandemic itself (iDeaths, iCases, iRt) are mostly depressing the value of the well-being up to summer, then some relief seems to appear. Clearly, some of these input are probably substitutes in terms of psychological effects. Another interesting aspect is the iUnemployment which seems to be relevant and negative before or at the onset of the pandemic (January and February) and before the start of the second wave (September), showing that, in the acute phase of the pandemic, increasing fears for health crowd out the economic concerns.

Looking at Japan, the variable jCases seems to be more important that jDeaths across months, probably because Japan experienced much lower number of fatalities than every other country in the world compared to the number of cases. In particular jDeaths shows a positive coefficient, which seems quite natural as the SWB-J decreases through time but the number of deaths is quite low an stable. Overall, for the whole period January to September, the news about coronavirus and COVID-19 seem to have a negative impact as well as the psychological health proxies (jStress, jHealth, jSolitude, jInsomnia), the mobility variables (jWorkplace, jResidential) and some of the economic variables (jGDPNews, jUnemployment). In more scattered way, this is confirmed through the months, where these variables appears to negatively impact the well-being in different times.

The reader may dig more into the analysis of the monthly regression and provide further interpretation of the coefficients and the impact of the different factors on the SWB indicators. Still, if we accept that the explanatory variables' impact on SWB rapidly changes over time, we should also accept that the calendar month cut may be a bit arbitrary. For this reason, we introduce a dynamic and more thorough approach, based on data science techniques, in the next section.

5.7.3 Dynamic Elastic Net Analysis

Having seen the features and limits of a yearly or monthly based analysis, we now present a dynamic model selection analysis. This means that instead of running a single fit of a statistical model on the whole or monthly data, we consider a sliding window of 30 days. In each 30 days-long time series data, we fit a the same statistical model discussed in the previous section, applying the Elastic Net approach (Zou and Hastie, 2005), which is a regularized estimation method that performs estimation and model selection at the same time.[18] In our setup, the Elastic Net method is a penalized least squares method which adds L_1 and L_2 penalization terms to the classical objective function of the OLS, i.e., it corresponds to the optimization problem in (5.7) below:

$$\operatorname*{argmin}_{\beta} \left\{ \frac{1}{2n} \sum_{i=1}^{n} (y_i - x_i'\beta)^2 + \lambda \left(\frac{1-\alpha}{2} \sum_{j=1}^{k} \beta_j^2 + \alpha \sum_{j=1}^{k} |\beta_j| \right) \right\}. \qquad (5.7)$$

where y_i is the dependent variable and $x_i = (x_{i1}, x_{i2}, \ldots, x_{ik})'$ is the vector of covariates for unit i in the sample of n observations, β_js are the regression coefficients, $\lambda > 0$ is a penalization factor and $\alpha \in [0,1]$ is a tuning parameter. For $\alpha = 1$ this method corresponds to the classical LASSO algorithm (Tibshirani, 1996), while for $\alpha = 0$ it corresponds to Ridge estimation (Hoerl and Kennard, 1970). Loosely speaking, while the LASSO method tends to estimate as zero as much coefficients as possible and, in case of multicollinearity, selects arbitrarily one single variable in a set of correlated covariates, the Ridge regression is able to accommodate for the multicollinearity by keeping the correlated variables and "averaging" the estimated coefficients. For exactly this reason Elastic Net include both L_1 (LASSO) and L_2 (Ridge) penalty terms. Usually LASSO is used to succinctly explain the correlation effects and Ridge is more suitable for forecasting. Overall, these methods are biased[19] but - with low variance, as they are also shrinkage methods, and hence most of the time - the Elastic Net performs quite well in terms of mean squared error compared to classical OLS. The value of $\alpha = 0.5$, usually denotes the proper Elastic Net. The penalty term λ is a tuning parameter and it is chosen through cross-validation methods. In this study we make use of the package **glmnet** developed by Friedman, Hastie, and Tibshirani (2010), which is computationally efficient and state of the art for this technique, to run several times the Elastic Net model. In particular, instead of a simple

---

[18]Contrary to the step-wise regression analysis, which is a hierarchical method that does not explore all possible solutions.

[19]Adaptive versions of Elastic Net also exist.

regression, we estimate a one step ahead forecasting model of this form

$$\operatorname*{argmin}_{\beta}\left\{\frac{1}{2\cdot 30}\sum_{d=t-29}^{t}(y_d - x'_{d-1}\beta)^2 + \lambda_t\left(\frac{1-\alpha}{2}\sum_{j=1}^{k}\beta_j^2 + \alpha\sum_{j=1}^{k}|\beta_j|\right)\right\},$$

$$(5.8)$$

for $t = 2020\text{-}12\text{-}02, \ldots$, where λ_t is calculated for each varying time t through cross-validation minimizing the mean squared error of the forecast. We call this version of Elastic Net the *Dynamic Elastic Net* approach (see also Carammia, Iacus, and Wilkin, 2020). In the set of

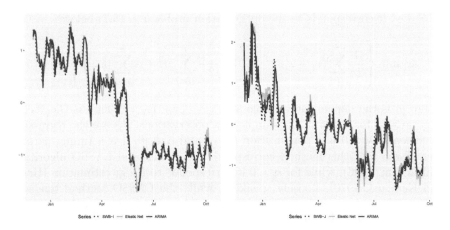

Figure 5.7
Forecast performance of Elastic Net and ARIMA(1,0,1). Standardized data.

regressors we also include the lagged value of the index (`swbLag`) and compare the forecasting against an ARIMA(1,0,1) model. We tested the average mean squared error of the forecast against that of the ARIMA(1,0,1) and they are quite close, with the latter being slightly better.[20] Clearly the ARIMA(1,0,1) model does not include any covariate but takes advantage of modeling the serial correlation, whilst Elastic Net in (5.8) essentially assumes independent observations - obviously a simplification of the reality - but it allows for the inclusion of explanatory variables. Figure 5.7 shows the relative performance of the Elastic Net compared to the ARIMA(1,0,1) model for both SWB-I and SWB-J. For completeness, we also tested the three values of $\alpha = 0, 0.5, 1$ getting that, as expected, $\alpha = 0.5$ gives the best performance. A final remark: as the real scope of this study is to evaluate the impact of the covariates on the SWB indexes, the actual forecasted values are not important, so we

---

[20]The average MSE is 0.01734827 for Elastic Net and 0.0152734 for ARIMA(1,0,1) for the Italian data and 0.08026962 and 0.06737153 respectively for the Japanese data.

standardize the data as Elastic Net works better with standardized data. Moreover, as the standard error of the estimates is quite difficult to obtain and it is not a typical outcome of the procedure, via the standardized data we also have standardized coefficients which make the evaluation of the impact of each variable easier to interpret (see Figures 5.8-5.9). Still, the relative importance of each variable is also obtained applying a Random Forests algorithm and extracting the importance measure from it. Once the importance measure is available, the covariates are ranked accordingly and a relative rank is calculated, where the relative rank is equal to 1 if the rank is highest, and is equal to 0 if the variable has not been selected (see Figures 5.10-5.11). Anyway, our main interest is in verifying the patterns of correlation through time. These patterns appears quite clearly indeed, as it will be discussed in the following.

5.7.4   Analysis of the Italian Data

By looking at Figure 5.8, it is clear that the variable iDeaths has a negative coefficient from mid February till the end of March (at the peak of the first wave), then again appears isolated on the 25th of May and again at the end of September and early October when the second waves restarted. Interestingly enough, iCases started much earlier around the end of January (remind that all flights from China to Italy were stopped on January 21st) and vanishes as iDeaths pops up. Then, its coefficient appears again and positive from mid May to mid June, when the good news of the decreasing in number of cases arrived. A similar pattern is exhibited by the iWuhan variable. It is worth noting that the iRt variable becomes negative by the end of April and quite persistent starting from the summer. This is clearly a media effect, as the notion of R_t itself became known to the public opinion only in July when both opinion makers and virologists started to discuss the indicator in the media, also on prime time TV shows.

The variable ilockdown and iDepression appear in mid March,[21] and the latter is also persistently negative between the end of April and mid May.

The Facebook survey variables FB.ILI and FB.CLI appear from their beginning (April 27th) and remain almost always negative (apart around the last weeks of July), meaning that the symptoms clearly impacted negatively the well-being.

The variable iUnemploymentNews is negative mostly from August, which is when the economic impact on several jobs, including seasonal jobs, had the largest impact and frequently reported in the news. It can be noted, moreover, that the iUnemployment effect is much less sharp, suggesting a higher incidence of the labor market perspectives, with respect to the impact of the current employment conditions.

---

[21]The lockdown in Italy started officially on March 15th.

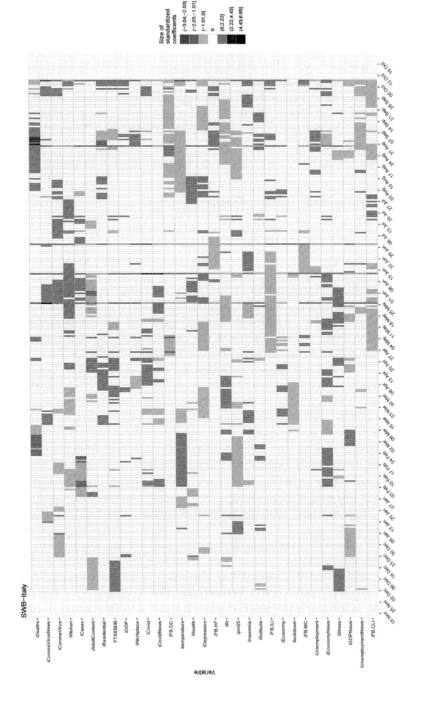

Figure 5.8
Standardized coefficients of the covariates selected by the Elastic Net for SWB-I through time.

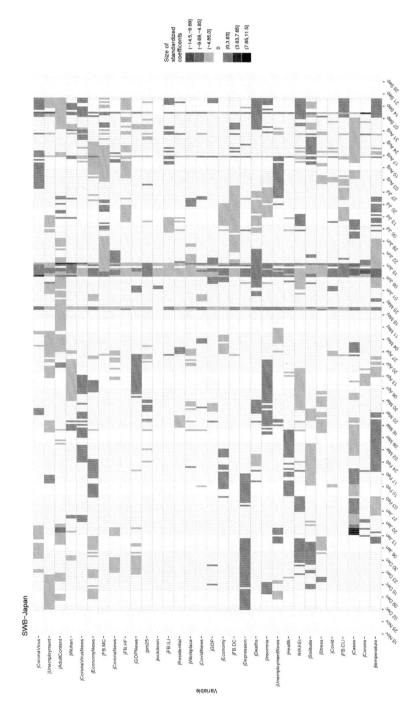

Figure 5.9
Standardized coefficients of the covariates selected by the Elastic Net for SWB-J through time.

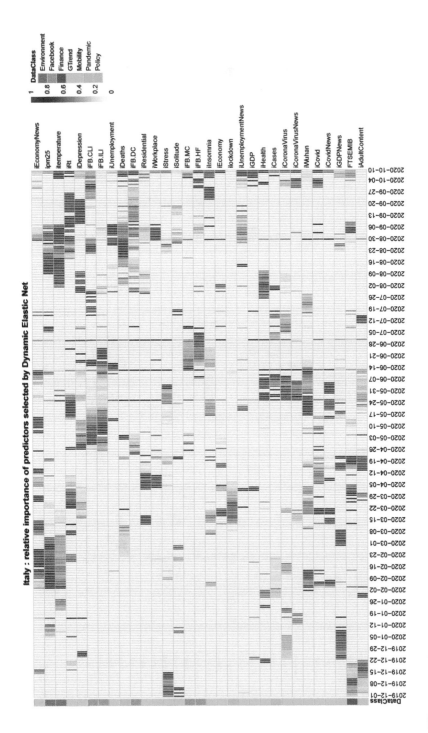

Figure 5.10
Relative importance (1 = maximum, 0 = variable not selected) of the covariates selected by Elastic Net to explain the SWB-I through time.

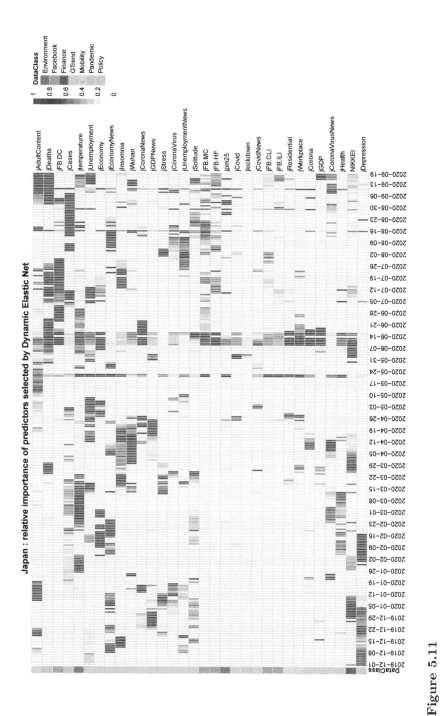

Figure 5.11
Relative importance (1 = maximum, 0 = variable not selected) of the covariates selected by elastic Net to explain the SWB-J through time.

The `itemperature` has a positive impact in the winter (which has been mild in Italy) and negative in the summer, which is somewhat confirmed by common experience, and the air pollution variable `ipm25` has almost always a negative impact when it is significant.

The mobility variables also show interesting patterns: `iResidential` is often positive unless in August, when staying at home was quite challenging, while `iWorkplace` has been negative around the last week of March and the first week of April, then generally positive when the mobility constrained where lifted in most regions.

The `iAdultContent` variable is always negatively correlated with the well-being, confirming that the consumption of these products are closely related to unsatisfactory levels of happiness.

The impact of economy related news and indexes are somehow questionable as they switch sign or are scattered along the time axis. This may indicate that these effects are extremely volatile and mainly determined by the correspondent news, as we may infer from `iGDP` and `iGDPNews`: in fact, while `iGDP` has sometimes a positive sign, the expected negative role seems played by `iGDPNews`. On the other hand, one may think that the impact of the economic factors is a long-run one, and hence is not properly captured by a short-run SWB indicator.

5.7.5   Analysis of the Japanese Data

The first wave of the pandemic started in Japan with the beginning of 2020, while - after a slowdown - a second wave arose in June. Figure 5.9 shows that the variable `jCases` is quite often present from January till May, then in June and August, and with almost always negative sign, whilst the `jDeaths` variable seems to be selected much less frequently and with alternate sign. This is probably because the number of deaths in Japan is way less than the number of cases. Also, the cases appeared in Japan much earlier than in Italy, and this is reflected by the selection of the variables.

Psychological aspects captured by the Google search `jStress` and `jSolitude` are are mostly present with negative sign while, surprisingly enough, `jInsomnia` and `jDepression` seems to be positively correlated with the SWB-J index, and the latter only up to mid February.

The economic variables - stock market index `NIKKEI`, `jEconomy`, `jFB.HF` and `jGDPNews` - are often selected and with negative impact on subjective well-being: in particular, labor market related covariates (`jUnemployment`, `jUnemploymentNews`) frequently appear, even with alternate sign, to measure concerns for current and future employment perspectives.

As expected, `jAdultContent` is negative when present, and seems to be relevant before the start of the pandemic and in between the two waves mostly. COVID-19 symptoms, as registered through the Facebook survey, seem to be less important apart for the period mid July to mid August when `jFB.CLI` appears with negative sign.

The mobility variables `jWorkplace`, `jResidential` and `jlockdown` appear with negative sign but they are not very persistent in time like it appeared for the Italian case. In fact, the mobility restrictions in Italy were much stronger and this may be the reason of this different impact. On the contrary, the social distancing proxies `jFB.MC` and `jFB.DC` appear and remain persistent during the second wave of the pandemic, and have both a negative impact on the SWB-J index.

The search terms related to Covid or coronavirus are also scattered and with alternate sign, showing a volatile relationship with on well-being, not so different from what one may observe in the Italian case.

The variable `jtemperature` behaves similarly to the Italian case: often positive in winter/spring and negative in June and August, and again positive in the first half of July. The air quality proxy `jpm2.5` is rarely significant compared to the Italian case.

Neither the postponement of the Olympic Games 2020, at the end of March, nor the consequent panic wave - with lines outside the supermarkets - both registered by Twitter and by SWB-J, are associated with the selection of specific covariates. The same happens with the resignation by the prime minister Shinzo Abe (end of August). On the other hand, the well-being reaction at the second wave outbreak in June is accompanied by a selection of all the covariates, some of which (`jCases`, `jCorona`) with a strongly negative impact.

5.7.6 Comparative Analysis of the Dynamic Elastic Net Results

Figure 5.12 summarizes the dynamic variable selection analysis for the two countries together. It shows the number of days each predictor has been selected by the elastic net model in equation (5.8) and the average relative importance of each variable as determined by the Random Forest algorithm after each step of the analysis. This representation is alternative to the one in Figures 5.10 and 5.11, as the time progression is not reported, but it still gives a glimpse of the overall impact of each variable along the whole period of analysis. In relative terms, it shows, for example, that the variables `jDeaths` and `jCases` are more frequently selected compared to the corresponding factors `iDeaths` and `iCases`, as well as the fact that the number of deaths is more important - compared to the number of cases - to explain SWB-I, while the number of cases is more important to explain SWB-J.

The `itemperature` and `jtemperature` variables are frequently selected in both countries, but this is not the case for the air quality proxy `ipm2.5` which is way more frequently selected than `jpm2.5` counterpart.

In general, the economic variables are more often selected in the Japanese case: see, for instance, `NIKKEI`, `jEconomyNews`, `jUnemployment` compared to the corresponding `FTSEMIB`, `iEconomyNews` and `iUnemployment` variables. On the contrary, `ilockdown` is more important compared to `jlockdown`, which is in line with the evidence reported in the previous sections and with the more stringent restrictive measures adopted in Italy. Further, the Covid-like

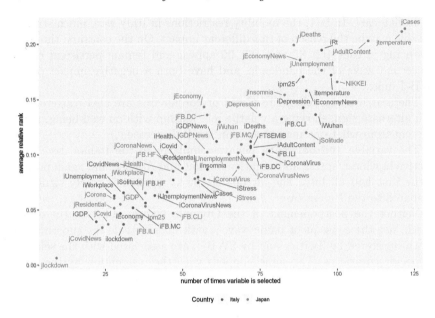

Figure 5.12
Summary of relative importance (average relative rank) and number of times
each variable is selected (over 315 dates for Italy and 294 dates for Japan) by
the dynamic elastic net.

symptoms variable `iFB.CLI` is more frequently selected than `jFB.CLI`, and
so forth.

All this to emphasize, once again, that the complexity of well-being determinants and their relationships - encompassing cultural dimensions, economic
status, environmental conditions as well as many contingent factors - is hardly
described by a single model.

5.8   Structural Equation Modeling

As seen in the previous sections, the relationships between the different factors
used in this study and the subjective well-being indexes are quite complex
and are mediated by an underlying latent variable that we may call subjective
well-being. Moreover, some of the covariates are clearly correlated and may
be considered as proxies of other latent dimensions. For these reasons, we try
to summarize these relationships via a Structural Equation Modeling (SEM)
as in Section 3.7.

Table 5.10

Ex-post interpretation of the latent variables of the SEM model. The `VirusSearch` and `Finance` convey similar information but with switched signs in both countries. The rest of the latent variables have the same and intuitive interpretation apart from `PsySearch` which has apparently no clear interpretation for the Italian case.

Latent	Country	Interpretation	Relationship
VirusSearch	IT	*pandemic getting better*	the higher its value, the lower the web search for reproduction number R_t and Covid in general
	JP	*fear about the pandemic*	the higher its value, the higher the web search for Corona virus and Covid in general
HealthStatus	IT, JP	*fear about symptoms*	the higher its value, the higher concerns about health
Mobility	IT, JP	*mobility restrictions*	the stronger its implementation, the harder to stand restrictions
SocDist	IT, JP	*social distancing practice*	the wider its use, the harder to stand distancing
Finance	IT	*fear for own economic conditions*	the higher its value, the higher pressure on own life
	JP	*confidence in country economy*	the higher its value, the better
PsySearch	IT	*search for psychological symptoms*	mixed evidence, hard to conclude
	JP	*psychological status*	the higher its value, the worse the well-being

We assume that well-being is a latent variable influenced by external factors. Further, the following latent dimensions, whose interpretation is postponed after the analysis of the results (see also Table 5.10), are assumed to exist:

- `VirusSearch`: captures the impact on `WellBeing` of Covid-related information web search. We assume the latent variable to be determined by the following observable covariates: `CoronaVirus`, `Covid` and, only for Italy by `Rt` and, only for Japan by `Corona`;

- `PsySearch`: captures the impact on `WellBeing` of web search for `Stress`, `Insomnia`, `Solitude` and `Depression`;

- `HealthStatus`: captures the declared health status, whose observables are `FB.CLI` and `FB.ILI`;

- `Mobility`: captures the mobility restrictions factor, measured through the variables `Residential`, `Workplace` and `lockdown`;

- `Finance`: accounts for the effect of the financial and economic variables: `FTSEMIB/NIKKEI`, `FB.HF` and `Unemployment`;

- `SocDist`: represents the social distancing factor, measured by `FB.MC` and `FB.DC`.

and the main causal relationships are as follows:

$$WellBeing \leftarrow VirusSearch + HealthStatus + Mobility$$
$$+ Finance + SocDist$$
$$PsySearch \leftarrow WellBeing$$
$$iAdultContent \leftarrow WellBeing$$
$$SWB\text{-}J/SWB\text{-}J \leftarrow WellBeing$$

Further, the residual correlations among some of the observed variables are inserted in the model to take into account cross-correlation, for example, between `Mobility` and `Cases`. The results of the fitted model are displayed in Tables 5.11–5.12, while Figures 5.13–5.14 give a graphical representation of the same fitting. In light of the results of the SEM models, Table 5.10 is an attempt to explain the meaning of the latent dimensions introduced in the analysis. The models have been fitted using the `lavaan` package (Rosseel, 2012) and plots have been generated through the `semPlot` package (Epskamp, 2019).

5.8.1   Evidence from the Structural Equation Modeling

Once established that the sign, the significance and the relative importance of observable variables in determining subjective well-being tend to rapidly change, we may try to derive from the SEM analysis a synthetic picture of the connections between the observables and well-being, with the help of latent dimensions.

Figures 5.13–5.14 show that the latent variables affect well-being with the expected sign, given the interpretation of these variables provided in Table 5.10: social distance measure and concerns, a worse confidence in economic and financial conditions, covid-like and flu symptoms, mobility restrictions and web search for Covid-related terms negatively impact well-being. We should note, however, that the `Finance` latent variable does not exhibit a significant effect on Japanese well-being, probably due to less severe concerns about the economic consequences of the pandemic, compared to the Italian case: we must remember that the estimated GDP reduction in 2020 in Japan is around a half with respect to what is expected in Italy.[22] Figure 5.15 shows the real

---

[22]The estimated real GDP growth in 2020 is -10.6% for Italy and -5.3% for Japan according

Table 5.11

The fitting results of the SEM model for the Italian data.

Relationship			Coefficient	Std. Err.
VirusSearch	↦	iCoronaVirus	0.226***	0.043
VirusSearch	↦	iCovid	−0.638***	0.063
VirusSearch	↦	iRt	−0.372***	0.057
PsySearch	↦	iStress	0.602***	0.053
PsySearch	↦	iInsomnia	0.410***	0.055
PsySearch	↦	iSolitude	1.003***	0.047
PsySearch	↦	iDepression	−0.186***	0.056
HealthStatus	↦	iFB.CLI	0.929***	0.047
HealthStatus	↦	iFB.ILI	1.010***	0.044
Mobility	↦	iResidential	0.791***	0.033
Mobility	↦	iWorkplace	−0.726***	0.039
Mobility	↦	ilockdown	0.762***	0.036
Finance	↦	FTSEMIB	−0.778***	0.052
Finance	↦	iFB.HF	0.521***	0.053
Finance	↦	iUnemployment	0.421***	0.051
SocDist	↦	iFB.MC	0.882***	0.047
SocDist	↦	iFB.DC	0.976***	0.044
WellBeing	↦	SWB-I	0.248***	0.019
WellBeing	↤	VirusSearch	0.307***	0.075
WellBeing	↤	HealthStatus	−0.167***	0.085
WellBeing	↤	Mobility	−0.290***	0.124
WellBeing	↤	Finance	−0.524***	0.101
WellBeing	↤	SocDist	−3.345***	0.336
PsySearch	↤	WellBeing	0.065***	0.016
iAdultContent	↤	WellBeing	−0.136***	0.016
PsySearch	cov	Mobility	0.003	0.008
Mobility	cov	SocDist	−0.412***	0.049
Mobility	cov	iCases	0.356***	0.034
VirusSearch	cov	Mobility	−0.293***	0.081
VirusSearch	cov	Finance	−1.227***	0.091
VirusSearch	cov	SocDist	−1.089***	0.072
HealthStatus	cov	Mobility	−0.002	0.057
HealthStatus	cov	Finance	0.230***	0.069
HealthStatus	cov	SocDist	0.314***	0.055
Mobility	cov	Finance	0.850***	0.033
Finance	cov	SocDist	0.305***	0.069
PsySearch	cov	iAdultContent	0.608***	0.046

Note: $^{*}p<0.1$; $^{**}p<0.05$; $^{***}p<0.01$

Table 5.12

The fitting results of the SEM model for the Japanese data.

Relationship			Coefficient	Std. Err.
VirusSearch	$\mapsto$	jCoronaVirus	0.867***	0.050
VirusSearch	$\mapsto$	jCovid	0.204***	0.061
VirusSearch	$\mapsto$	jCorona	0.714***	0.054
PsySearch	$\mapsto$	jStress	0.401***	0.055
PsySearch	$\mapsto$	jInsomnia	0.749***	0.049
PsySearch	$\mapsto$	jSolitude	0.763***	0.049
PsySearch	$\mapsto$	jDepression	0.545***	0.052
HealthStatus	$\mapsto$	jFB.CLI	1.016***	0.042
HealthStatus	$\mapsto$	jFB.ILI	0.947***	0.045
Mobility	$\mapsto$	jResidential	1.071***	0.039
Mobility	$\mapsto$	jWorkplace	−0.902***	0.046
Mobility	$\mapsto$	jlockdown	0.708***	0.050
Finance	$\mapsto$	NIKKEI	0.836***	0.050
Finance	$\mapsto$	jFB.HF	−0.441***	0.054
Finance	$\mapsto$	jUnemployment	−0.694***	0.053
SocDist	$\mapsto$	jFB.MC	0.959***	0.044
SocDist	$\mapsto$	jFB.DC	0.903***	0.047
WellBeing	$\mapsto$	SWB-J	0.674***	0.036
WellBeing	$\hookleftarrow$	VirusSearch	−0.368***	0.134
WellBeing	$\hookleftarrow$	HealthStatus	−0.617***	0.240
WellBeing	$\hookleftarrow$	Mobility	−0.077	0.121
WellBeing	$\hookleftarrow$	Finance	0.052	0.165
WellBeing	$\hookleftarrow$	SocDist	−1.795***	0.298
PsySearch	$\hookleftarrow$	WellBeing	−0.342***	0.052
jAdultContent	$\hookleftarrow$	WellBeing	−0.456***	0.039
PsySearch	cov	Mobility	0.002	0.023
Mobility	cov	SocDist	0.087	0.054
Mobility	cov	jCases	0.235***	0.030
VirusSearch	cov	HealthStatus	0.536***	0.049
VirusSearch	cov	Mobility	0.192***	0.056
VirusSearch	cov	Finance	−0.947***	0.031
VirusSearch	cov	SocDist	−0.566***	0.051
HealthStatus	cov	Mobility	−0.278***	0.047
HealthStatus	cov	Finance	−0.269***	0.062
HealthStatus	cov	SocDist	−0.928***	0.010
Mobility	cov	Finance	−0.676***	0.034
Finance	cov	SocDist	0.450***	0.059
PsySearch	cov	jAdultContent	0.247***	0.047

Note: $^*p<0.1$; $^{**}p<0.05$; $^{***}p<0.01$

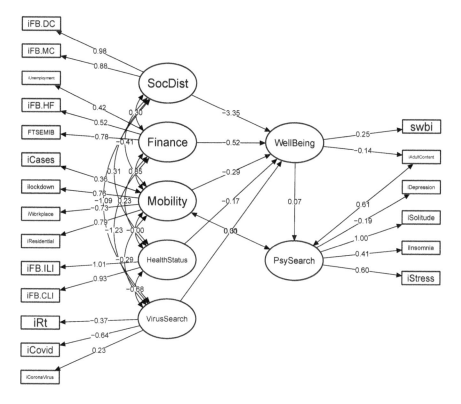

Figure 5.13
Output of the SEM fitting for SWB-I.

GDP growth in 2020 and the number of COVID-19 deaths per 1 million inhabitants, and it is quite informative of the relative impact of COVID-19 on the different economies around the world.

A partial exception is represented by the latent variable `PsySearch`, that it is supposed to summarize the psychological status measured by the web search of terms related to anxiety, loneliness and depression: while this variable is negative impacted by well-being in Japan, this does not happen for the Italian case, where the relationship is surprisingly positive and significant: this is one of the cases where cultural habits and social norms may play a role in producing different results in different countries.

Finally, we have to point out the positive correlation between the two SWB indexes and the corresponding well-being latent variable. On one hand, this pleads for the ability of the indicators to read into public opinion; on the other hand, we cannot help noting that the relationship is much stronger for

to the International Monetary Fund. The full report can be found at: `https://www.imf.org/external/datamapper/NGDP_RPCH@WEO/OEMDC/ADVEC/WEOWORLD?year=2020`

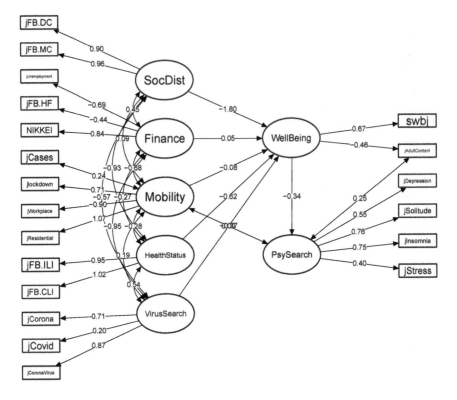

Figure 5.14
Output of the SEM fitting for SWB-J.

SWB-J, suggesting that the Twitter-based tool better capture the individual and collective sentiment in the Japanese context.

5.9 Summary of the Results

The main evidence of this study is that a single approach cannot give a full account about the complex relationship among subjective well-being and the wide range of potential explanatory variables, when an unprecedented shock hits the socio-economic context. Nevertheless, the multi-method analysis conducted in the previous sections let a few important evidences emerge:

- the Twitter indexes SWB-I and SWB-J show that Italy and Japan have a substantial difference in their own self-perceived well-being: the Italian indicator is permanently higher than the Japanese one. On the other hand, both SWB-I and SWB-J report a substantial decrease in 2020 compared to the previous years (-11.7% for Italy and -8.3% for Japan);

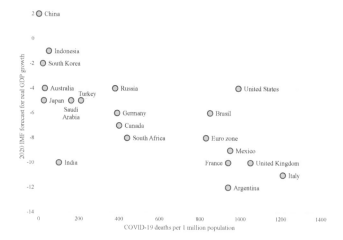

Figure 5.15
Number of COVID-19 deaths (WHO) and real GDP forecast growth in 2020 (IMF).

- applying a stochastic differential equations analysis, it can be seen that this drop has stabilized since April/May for both countries, meaning that the pandemic shock is provoking a prolonged status of emergency and stress, with permanent negative impact on the self-perceived well-being trend;

- the stochastic analysis also allows for disentangling long term (or structural) and short term (or emotional) components of the subjective well-being, showing that, if the well-being decrease in 2020 is higher in Italy, the convergence to the new long term well-being level is faster in Japan. Implicitly, this analysis clarifies why yearly survey may lead to biased evidence of the real well-being level of a country;

- both the SWB-I and SWB-J Twitter indexes, despite the cultural differences of the two countries, seems to be able to capture some aspects of subjective well-being and the relationships with potential explanatory variables. This ability is not affected by language differences and specificities but, not surprisingly, requires a dynamic analysis to emerge: one of the main advantages of a Twitter-based index, in fact, is in its high frequency and in the opportunity to capture even rapid changes in the relative importance of determinants;

- in fact, the Dynamic Elastic Net analysis has captured the complexity of the well-being determination: which factors matter more than others and when. Although a general model of well-being is desirable, the analysis has shown that according to different context conditions (pre-pandemic, pandemic, in-between waves of pandemic, etc.) some determinants weight more than others and they interact differently with cultural aspects in the two countries;

- even if we think to have documented that "average" effects may be misleading guides in interpreting dynamic phenomena, the structural equation modeling analysis has shown some common latent dimensions that can help in describing a causal model for the latent variable of subjective well-being of which the Twitter indexes SWB-I and SWB-J seem to be measurable proxies. In particular, social distances concerns, mobility restrictions, economic conditions, along with more obvious health conditions and Covid-related concerns, have been identified as possible explanatory variables. At the same time, variables linked to psychological dimensions show less expected effects, indicating the need for a more thorough understanding of cultural specificities of the examined countries.

5.10 Conclusions

We have seen how different methods to extract well-being produce indexes that react differently to the CODIV-19 pandemic. In practice, we have seen that both the Hedonometer and the Gross National Happiness Index have negative peaks around the dates of the declaration of emergency state, and then the indicators go back to the previous pre-pandemic level more or less fast. On the contrary, SWB-J and SWB-I seem to show that the well-being decreases and keep a low level till the end of the summer. The World Well-being Project show similarly that the pandemic have shifted the use of words related to mental and physical health remarkably and permanently, after the lockdown measures have been put in place.

Despite the noticeable differences in mean reversion processes, however, all the approaches show that lockdown and social distancing measures play a role in reducing the overall happiness from 2019 to 2020.

The Dynamic Elastic Net approach have also shown that the well-being process is indeed complex and the correlation between the SWB indexes and several other possible explanatory variables do change over time, meaning that well-being moves around its structural component, following the shifts induced by a wide range of factors.

All the analysis, by all methods, show that a high frequency indicator of well-being is necessary to better understand the dynamics of the well-being itself.

5.11 Glossary

Dynamic Elastic Net: Elastic Net method of a time varying window.

Elastic Net: Estimation and model selection regularization method.

SEM: Structural equation model.

Bibliography

S. Abdullah, E. L. Murnane, J. M. R. Costa, and T. Choudhury. "Collective Smile: Measuring Societal Happiness from Geolocated Images". In: *Proceedings of the 18th ACM Conference on Computer Supported Cooperative Work & Social Computing*. CSCW '15. ACM, 2015, pp. 361–374.

M. Alexander, K. Polimis, and E. Zagheni *Combining social media and survey data to nowcast migrant stocks in the United States*. 2020. arXiv: 2003.02895 [stat.AP].

Y. Algan, E. Beasley, F. Guyot, K. Higa, F. Murtin, and C. Senik. *Big Data Measures of Well-Being: Evidence from a Google Well-Being Index in the United States*. Tech. rep. 3. OECD Statistics Working Papers, 2016.

T. Alshaabi, J. R. Minot, M. V. Arnold, J. L. Adams, D. R. Dewhurst, A. J. Reagan, R. Muhamad, C. M. Danforth, and P. S. Dodds. *How the world's collective attention is being paid to a pan- demic: COVID-19 related n-gram time series for 24 languages on Twitter*. 2020. arXiv: 2003.12614[physics.soc-ph].

J.A. Anderson. *Introduction to Neural Networks*. Cambridge, MA: MIT Press, 1995.

H. Asaoka, Y. Koido, Y. Kawashima, M. Ikeda, Y. Miyamoto, and D. Nishi. "Post-traumatic stress symptoms among medical rescue work- ers exposed to COVID-19 in Japan". In: *Psychiatry and Clinical Neuro- sciences* 74.9 (2020), pp. 503–505.

A. Asuncion, M. Welling, P. Smyth, and Y. Teh. "On smoothing and inference for topic models". In: *Uncer- tainty in Artificial Intelligence* (2009).

R. Baker, J. M. Brick, N. A. Bates, M. Battaglia, M. P. Couper, J. A. Dever, K. J. Gile, and R. Tourangeau. "Summary Report of the AAPOR Task Force on Non- probability Sampling". In: *Journal of Survey Statistics and Methodology* 1.2 (2013). pp. 90–143.

R. Baldwin and B. Weder di Mauro. *Economics in the Time of Covid-19*. CEPR Press, 2020.

J. Barameechai. "The Green and Happiness Index". In: *International Conference on Happiness and Public Policy*. Bangkok, 2007.

C. Barrington-Leigh and A. Escande. "Measuring progress and well-being: A comparative review of indicators". In: *Social Indicators Research* 135.3 (2018), pp. 893–925.

N. K. Baym. *Personal Connections in the Digital Age.* 2nd edition. Digital Media and Society Series. Polity Press, 2015.

A. Ben-Hur, D. Horn, H. Siegelmann, and V.N Vapnik. "Support vector clustering". In: *Journal of Machine Learn- ing Research* 2 (2001), pp. 125–137.

Y. Bengio, R. Ducharme, P. Vincent, and C. Janvin. "A neural probabilistic language model". In: *The Journal of Machine Learning Research* 3 (2003), pp. 1137–1155.

D. J. Benjamin, K. B. Cooper, O. Heffetz, and M. S. Kimball. "Self-reported wellbeing indicators are a valuable complement to traditional economic indicators but are not yet ready to compete with them". In: *Behavioural Public Policy* 4.2 (2020), pp. 198– 209.

BETI. *Coronavirus causes massive declines in the BETI Economic transactions declined to 2008 levels.* 2020. `https://www.bankservafrica.com/blog/`.

F. Black and M.S. Scholes. "The pricing of options and corporate liabilities". In: *Journal of Political Economy.* 81 (1973), pp. 637–665.

D.G. Blanchflower and A.J. Oswald. "Well-being over time in Britain and the USA". In: *Journal of Public Economics* 88.7–8 (2004), pp. 1359–1386.

D.M Blei and J. Lafferty. "A correlated topic model of science". In: *The Annals of Applied Statistics* 1.1 (2007), pp. 17–35.

D. M. Blei and J. Lafferty. "Dynamic topic models". In: *International Conference on Machine Learning* (2006), pp. 113–120.

D.M. Blei. "Probabilistic Topic Models". In: *Communications of ACM* 55.4 (2012), pp. 77–84.

D.M. Blei, A. Ng, and M. Jordan. "Latent Dirichlet allocation". In: *Journal of Machine Learning Research* 3 (2003), pp. 993–1022.

D. M. Blei and J. D. McAuliffe. "Supervised Topic Models". In: *Pro- ceedings of the 20th International Conference on Neural Information Pro- cessing Systems.* NIPS'07. USA: Curran Associates Inc., 2007, pp. 121–128.

C. A. Bliss, I. M. Kloumann, K. D. Harris, C. M. Danforth, and P. S. Dodds "Twitter reciprocal reply networks exhibit assorta- tivity with respect to happiness". In: *CoRR* (2011). `http://arxiv.org/abs/1112.1010`.

P. de Boeck and M. Wilson. *Explanatory Item Response Models: A Generalized Linear and Nonlinear Approach.* New York: Springer, 2004.

J. Bollen, B. Gonçalves, I. van de Leemput, and G. Ruan. "The happiness paradox: Your friends are happier than you". In: *EPJ Data Science* 6.4 (2017), pp. 1–10.

J. Bollen, B. Gonçalves, G. Ruan, and H. Mao. "Happiness is as- sortative in online social networks". In: *Artificial Life* 17.3 (Aug. 2011), pp. 237–251.

K.A. Bollen. *Structural Equations with Latent Variables.* New York: Wiley, 1989.

M. Bouchet-Valat *SnowballC: Snowball stemmers based on the C libstemmer UTF-8 library.* 2014. http://CRAN.R-project.org/package= SnowballC. R package version 0.5.1.

L. Breiman. "Random forests". In: *Machine Learning* 45.1 (2001), pp. 5–32.

L. Breiman, J. H. Friedman, R. A. Olshen, and C. J. Stone. *Classification and Regression Trees.* Monterey: Wadsworth and Brooks-Cole, 1984.

A. Brodeur, A. E. Clark, S. Fleche, and N. Powdthavee. *Assessing the impact of the coronavirus lockdown on unhappiness, loneliness, and boredom using Google Trends.* 2020. arXiv: 2004.12129 [physics.soc-ph].

A. Brouste, M. Fukasawa, H. Hino, S. M. Iacus, K. Kamatani, Y. Koike, H. Masuda, R. Nomura, T. Ogihara, Y. Shimuzu, M. Uchida, and N. Yoshida. "The YUIMA Project: A computational framework for simulation and inference of stochastic differential equations". In: *Journal of Statistical Software* 57.4 (2014), pp. 1–51.

M. N. Burns, M. Begale, J. Duffecy, D. Gergle, C. J. Karr, E. Giangrande, and D. C. Mohr. "Harnessing context sensing to develop a mobile intervention for depression". In: *Journal of Medical Internet Research* 13.3 (2011), pp. 158–174.

Centre for Buthan Studies & GNH Research. *Gross National Index.* Tech. rep. Centre for Buthan Studies, 2015. http://www.grossnationalhappiness.com/articles/.

R. Capuano, M. Altieri, A. Bisecco, A. d'Ambrosio, R. Docimo, D. Buonanno, F. Matrone, F. Giuliano, G. Tedeschi, G. Santangelo, and A. Gallo. "Psychological consequences of COVID-19 pandemic in Italian MS patients: Signs of resilience?" In: *Journal of Neurology* July.28 (2020), pp. 1–8.

M. Carammia, S. M. Iacus, and T. Wilkin. *Forecasting asylum-related migration flows with machine learning and data at scale.* 2020. arXiv: 2011.04348[stat.AP].

T. Carpi and S. M. Iacus. "Is Japanese gendered language used on Twitter? A large scale study". In: *Online Journal of Communication and Media Technologies* 10.4 (2020), e202024.

A. Case and A. Deaton. "Suicide, age, and wellbeing: An empirical investigation". In: *NBER Working Paper* (2015), pp. 1–42.

A. Ceron, L. Curini, and S. M. Iacus. *Social Media e Sentiment Analysis: L'evoluzione dei fenomeni sociali attraverso la Rete*. Sxl - Springer for Innovation, Springer, 2013.

A. Ceron, L. Curini, and S. M. Iacus. "Using sentiment analysis to monitor electoral campaigns: Method matters? Evidence from the United States and Italy". In: *Social Science Computer Review* 33.1 (2015), pp. 3–20.

A. Ceron, L. Curini, and S. M. Iacus. "iSA: A fast, scalable and accurate algorithm for sentiment analysis of social media content". In: *Information Sciences* 367–368 (2016), pp. 105–124.

N. Cesare, H. Lee, T. McCormick, E. Spiro, and E. Zaghen. "Promises and pitfalls of using Digital traces for demo- graphic research". In: *Demography* 55.5 (2018), pp. 1979–1999.

K. C. Chan, G. A. Karolyi, F. A. Longstaff, and A. B. Sanders. "An empirical investigation of alternative models of the short-term interest rate". In: *Journal of Finance* 47 (1992), pp. 1209–1227.

J. Chang and D. M. Blei. "Hierarchical relational models for document networks". In: *Annals of Applied Ststistics* 4.1 (2010), pp. 124–150.

H. Choi and H. Varian. "Predicting the present with Google Trends". In: *Economic Record* 88.s1 (2012), pp. 2–9.

M. A. Choudhary, P. Levine, P. McAdam, and P. Welz. "The happiness puzzle: Analytical aspects of the Easterlin paradox". In: *Oxford Economic Papers* 64.1 (2012), pp. 27–42.

A. Chudik, K. Mohaddes, M. H. Pesaran, M. Raissi, and A. Rebucci. "Economic consequences of Covid-19: A counterfactual multi- country analysis". 2020. `https://voxeu.org/print/66360`.

A.E. Clark and A.J. Oswald. "Unhappiness and unemployment". In: *Economic Journal* 104.424 (1994), pp. 648–659.

A. Clark *Inequality-Aversion and Income Mobility: A Direct Test*. DELTA Working Papers 2003-11. DELTA (Ecole normale supérieure), 2003.

L.A. Clark and D. Pregibon. "Tree-based Models". In: *Statistical Models in S*. Ed. by J.M. Chamber and T.J. Hastie. New York: Chapman & Hall, 1992.

D. Cooper and M. Greenaway. *Non-probability Survey Sampling in Offi- cial Statistics*. Office for National Statistics - Methodology Working Paper Series N.4, 2015.

C. Cortes and V.N. Vapnik. "Random forests". In: *Machine Learning* 20.3 (1995), pp. 273–297.

M.P. Couper. "Is the sky falling? New technology, changing media, and the future of surverys". In: *Survey Research Methods* 7.3 (2013), pp. 145–156.

J.C. Cox, J.E. Ingersoll, and S.A. Ross. "A theory of the term structure of interest rates". In: *Econometrica* 53 (1985), pp. 385–408.

N. Cristianini and J. Shawe-Taylor. *An Introduction to Support Vector Machines and Other Kernel-based Learning Methods.* Cambridge: Cambridge University Press, 2000.

F. Cugnata, S. Salini, and E. Siletti. "Measuring well-being com- bining different data sources: A Bayesian networks approach". In: *Società Italiana di Statistica. SIS 2020.* 2020, pp. 685–690.

L. Curini, S.M. Iacus, and L. Canova. "Measuring idiosyncratic happiness through the analysis of Twitter: An application to the Italian case". In: *Social Indicators Research* 121.2 (2015), pp. 525–542.

F. D'Orlando. "The demand for pornography". In: *Journal of Happiness Studies* 12.1 (2011), pp. 51–75.

K. Dave, S. Lawrence, and D.M. Pennock. "Mining the peanut gallery: Opinion extraction and semantic classification of product reviews". In: *Proceedings of WWW 2003.* 2003.

A. Deaton. "The financial crisis and the wellbeing of Americans". In: *Ox- ford Economic Papers* 64.1 (2012), pp. 1–26.

A. Deaton and A. A. Stone. "Two Happiness Puzzles." *American Eco- nomic Review,* 103.3: 591–597. 2013.

A. Deaton and A. A. Stone. "Understanding context effects for a measure of life evaluation: How responses matter". In: *Oxford Economic Papers* 68.4 (2016), pp. 861–870.

P. Deb, D. Furceri, J. D. Ostry, and N. Tawk. "The Economic Effects of Covid-19 Containment Measures". 2020. https://www.imf.org/en/Publications/WP/Issues/2020/08/07/The-Economic-Effects-of-COVID-19-Containment-Measures-49571.

R. Di Tella and R. MacCulloch. "Gross national happiness as an answer to the Easterlin Paradox?" In: *Journal of Development Economics* 86.1 (2008), pp. 22–42.

E. Diener "Subjective well-being". In: *Psychological Bulletin* 95.3 (1984), pp. 542–563.

E. Diener "Guidelines for national indicators of subjective well-being and ill-being". In: *Journal of Happiness Studies* 7.4 (2006), pp. 397–404.

E. Diener, R.A. Emmons, R.J. Larsen, and S. Griffin. "The satisfaction with life scale". In: *Journal of Personality Assessment* 49.1 (1985), pp. 71–75.

E. Diener, W. Ng, J. Harter, and R. Arora. "Wealth and happiness across the world: Ma- terial prosperity predicts life evaluation, whereas psychosocial prosperity predicts positive feeling." In: *Journal of Personality and Social Psychology* 99.1 (2010), pp. 52–61.

E. Diener, E. M. Suh, R. E. Lucas, and H. L. Smith. "Subjective well-being: Three decades of progress". In: *Psychological Bulletin* 125.2 (1999), pp. 276– 302.

E. Diener, E. M. Suh, H. L. Smith, and L. Shao. "National differences in reported subjective well-being: Why do they occur?" In: *Social Indicators Research* 34.1 (1995), pp. 7–32.

P. S. Dodds and C. M. Danforth. "Measuring the happi- ness of large-scale written expression: Songs, blogs, and presidents". In: *Journal of Happiness Studies* 11.4 (2010), pp. 441–456.

P. S. Dodds, K. D. Harris, I. M. Kloumann, C. A. Bliss, and C. M. Danforth. "Temporal patterns of happiness and information in a global social network: Hedonometrics and Twitter". In: *PLoS ONE* 6.12 (2011).

N. Döring. "How is the COVID-19 pandemic affecting our sexualities? An overview of the current media narratives and research hypotheses". In: *Archives of Sexual Behavior* 49.8 (2020), pp. 2765–2778.

A. Dubois, E. Zagheni, K. Garimella, and I. Weber. "Studying Migrant Assimilation Through Facebook Inter- ests". In: *Lecture Notes in Computer Science, Social Informatics*. Ed. by S. Staab, O. Koltsova, and D. Ignatov. Vol. 11186. Cham: Springer, 2018, pp. 51–60.

A. O. Durahim and M. Coşkun. "#iamhappybecause: Gross Na- tional Happiness through Twitter analysis and big data". In: *Technologi- cal Forecasting and Social Change* 99 (2015), pp. 92–105.

R. A. Easterlin. "Does Economic Growth Improve the Human Lot?" In: *Nations and Households in Economic Growth: Essays in Honor of Moses Abramovitz*. Ed. by Paul A. David and Melvin W. Reder. Academic Press, 1974, pp. 89–125.

Israeli Ministry of Environmental Protection. *Well-being Indicators for Israel*. Tech. rep. Israeli Ministry of Environmental Protection, 2014. http://www.sviva.gov.il/English/Indicators/Documents/Well-Being-Indicators-for-Israel-April2014.pdf .

S. Epskamp. *semPlot: Path Diagrams and Visual Analysis of Various SEM Packages' Output*. CRAN. 2019. `https://CRAN.R-project.org/package=semPlot\%7D`. R package version 1.1.2.

Eurostat. *Population on 1 January by Age Group, Sex and Citizenship*. Available: `https://appsso.eurostat.ec.europa.eu/nui/show.do?dataset=migr\_pop1ctz`. Jan. 2018.

Facebook. *Facebook Ads Manager*. Available: `https://www.facebook.com/adsmanager/creation`. Mar. 2020.

S. Falorsi, A. Fasulo, A. Naccarato, and M. Pratesi. *Small Area model for Italian regional monthly estimates of young unemployed using Google Trends Data*. 61*st* World Congress of the International Statistical Institute, 16–21 July 2017, Marrakech (MA). Oct. 2017.

J. Fan, F. Han, and H. Liu. "Challenges of Big Data analysis". In: *National Science Review* 1.2 (2014), pp. 293–314.

J. Fan, Y. Li, K. Stewart, A. R. Kommareddy, A. Bradford, S. Chiu, F. Kreuter, N. Barkay, A. Bilinski, B. Kim, R. Eliat, T. Galili, D. Haimovich, S. LaRocca, S. Presser, K. Morris, J. A. Salomon, E. A. Stuart, R. Tibshirani, T. A. Barash, C. Cobb, A. Garcia, A. Gros, A. Isa, A: Kaess, F. Karim, O. E. Kedosha, S. Matskel, R. Melamed, A. Patankar, I. Rutenberg, T. Salmona, and D. Vannette. "COVID-19 World Symptom Survey Data API". 2020. `https://covidmap.umd.edu/api.html`.

R. E. Fay and R. A. Herriot. "Estimates of income for small places: An application of James-Stein procedures to census data". In: *Journal of the American Statistical Association* 74.366 (1979), pp. 269–277.

R. Feldman and S. James. *The Text Mining Handbook*. New York: Cambridge University Press, 2007.

N. Fernandes. "Economic Effects of Coronavirus Outbreak (COVID-19) on the World Economy". In: *IESE Business School Working Paper No. WP-1240-E* E.1240 (2020).

R.M. Filho, J.M. Almeida, and G.L. Pappa. "Twitter population sample bias and its impact on predictive outcomes: A case study on elections". In: *2015 IEEE/ACM International Conference on Advances in Social Networks Analysis and Mining (ASONAM)*. 2015, pp. 1254–1261.

M. Fleurbaey. "Beyond GDP: The quest for a measure of social welfare". In: *Journal of Economic Literature* 47.4 (2009), pp. 1029–1075.

R. Foa, S. Gilbert, and M. O. Fabian. "Covid-19 and Subjective Well-Being: Separating the Effects of Lockdowns from the Pandemic". 2020. `https://www.bennettinstitute.cam.ac.uk/media/uploads/files/Happiness\_under\_Lockdown.pdf`.

M.T. Ford, A.T. Jebb, L. Tay, and E. Diener. "Internet searches for affect-related terms: An indicator of subjective well-being and predictor of health outcomes across US States and metro areas." In: *Applied Psychology: Health and Well-Being* 10.1 (2018), pp. 3–29.

M. J. C. Forgeard, E. Jayawickreme, M. Kern, and M. E. P. Seligman. "Doing the right thing: Measuring wellbeing for public policy". In: *International Journal of Wellbeing* 1.1 (2010), pp. 79–106.

F. Fors and J. Kulin. "Bringing affect back in: Measuring and com- paring subjective well-being across countries". English. In: *Social Indica- tors Research* 127.1 (2016), pp. 323–339.

New Economics Foundation. *The Happy Planet Index: 2012 Report. A Global Index of Sustainable Well-being.* Tech. rep. New Economics Foundation, 2012. http://www.happyplanetindex.org/assets/happy-planet-index-report.pdf.

M. R. Frank, L. Mitchell, P. S. Dodds, and C. M. Danforth. "Happiness and the patterns of life: A study of geolocated tweets". In: *Scientific Reports* 3, 2625 (2013).

B. S. Frey and A. Stutzer. "What can economists learn from happi- ness research?" In: *Journal of Economic Literature* 40.2 (2002), pp. 402–435.

J. Friedman, T. Hastie, and R. Tibshirani. "Regularization paths for generalized linear models via coordinate descent". In: *Journal of Statistical Software* 33.1 (2010), pp. 1–22.

P. Frijters, A. E. Clark, C. Krekel, and R. Layard. "A happy choice: Wellbeing as the goal of government". In: *Behavioural Public Policy* 4.2 (2020), pp. 126–165.

K. Fukushima. "Neural network model for a mechanism of pattern recognition unaffected by shift in position - Neocognitron". In: *Trans. IECE* J62-A.10 (1979), pp. 658–665.

K. Fukushima. "Neocognitron: A self-organizing neural network for a mechanism of pattern recognition unaffected by shift in position". In: *Biological Cybernetics* 36.4 (1980), pp. 193–202.

K. Fukushima. "Artificial vision by multi-layered neural networks: Neocognitron and its advances". In: *Neural Networks* 37 (2013), pp. 103–119.

A. Gelman, D. Park, B. Shor, J. Bafumi, and J. Cortina. *Red State, Blue State, Rich State, Poor State.* Princeton: Princeton University Press, 2008.

M. Ghosh, N. Nangia, and D. H. Kim. "Estimation of median in- come of four-person families: A Bayesian time series approach". In: *Journal of the American Statistical Association* 91.436 (1996), pp. 1423–1431.

M. Giesselmann, R. Hilmer, N. A. Siegel, and G. G. Wagner. "Measuring well-being: W3 indicators to comple- ment GDP". In: *DIW Economic Bulletin* 3.5 (2013), pp. 10–19.

A.D. Gordon. *Classification, 2nd Ed.* Boca Raton, FL: Chapman & Hall/CRC, 1999.

F. Greco and A. Polli. "Security perception and people well-being". In: *Social Indicators Research* 153 (2021). pp. 741–758.

E. Grewenig, P. Lergetporer, L. Simon, K. Werner, and L. Woessmann. *Can Online Surveys Represent the Entire Popula- tion?* Tech. rep. 11799. IZA Discussion Paper, 2018.

T. Greyling, S. Rossouw, and T. Adhikari *Happiness-lost: Did Governments Make the Right decisions to Combat Covid-19?* GLO Discussion Paper 556. Global Labor Organization, 2020. http://hdl.handle.net/10419/217494.

T. L. Griffiths and M. Steyvers. "A Probabilistic Approach to Semantic Rep- resentation". In: *Proceedings of the 24th Annual Conference of the Cognitive Science Society.* 2002.

J. Grimmer and G. King. "General purpose computer-assisted clustering and conceptualization". In: *Proceedings of the National Academy of Sciences* 108.7 (2011), pp. 2643–2650.

J. Grimmer and B. M. Stewart. "Text as data: The promise and pitfalls of automatic content analysis methods for political texts". In: *Political Analysis* 21.3 (2013), pp. 267–297.

M. R. Gualano, G. Lo Moro, G. Voglino, F. Bert, and R. Siliquini. "Effects of Covid-19 lockdown on mental health and sleep disturbances in Italy". In: *International Journal of Environmental Research and Public Health* 17.3 (2020), pp. 1–13.

S. C. Guntuku, G. Sherman, D. C. Stokes, A. K. Agarwal, E. Seltzer, R. M. Merchant, and L. H. Ungar. "Tracking mental health and symptom men- tions on Twitter during COVID-19". In: *Journal of General Internal Medicine* 35.9 (2020), pp. 2798–2800.

R.H Hall. "Internet Use and Happiness". In: *HCI in Business, Government, and Organizations: eCommerce and Innovation.* Ed. by F.H. Nah and C.H. Tan. Vol. 9751. Lecture Notes in Computer Science. Cham: Springer, 2016.

R.H. Hall. "Internet Use and Happiness: A Longitudinal Analysis". In: *HCI in Business, Government and Organizations. Supporting Business*. Ed. by F.H. Nah and C.H. Tan. Vol. 10294. Lecture Notes in Computer Science. Cham: Springer, 2017.

E. Hargittai. "Potential biases in Big Data: Omitted voices on social media". In: *Social Science Computer Review* 38.1 (2020), pp. 10–24.

M. Haruna and D. Nishi. "Perinatal mental health and COVID-19 in Japan". In: *Psychiatry and Clinical Neurosciences* 74.9 (2020), pp. 502–503.

T. Hastie, R. Tibshirani, and J. Friedman. *The Elements of Statistical Learning, 2nd ed.* New York: Springer, 2008.

W.K. Hastings. "Monte carlo sampling methods using Markov chains and their applications". In: *Biometrika* 57 (1970), pp. 97–109.

S. Haykin. *Neural Networks: A Comprehensive Foundation, 2nd ed.* Englewood Cliffs, NJ: Prentice-Hall, 1999.

J.J. Heckman. "Sample selection bias as a specification error". In: *Econometrica* 47.1 (1979), pp. 153–161.

C. R. Henderson. "Best linear unbiased estimation and prediction under a selection model". In: *Biometrics* 31.2 (1975), pp. 423–447.

A. Hino and R. A. Fahey. "Representing the Twittersphere: Archiving a representative sample of Twitter data under resource con- straints". In: *International Journal of Information Management* 48 (2019), pp. 175–184.

S. Hochreiter, A.S. Younger, and P.R. Conwell. "Learning to learn using gradient descent". In: *Proceedings of the International Conference on Artificial Neural Networks (ICANN 2001)*. Vol. In Lecture Notes on Computer Science 2130. Springer: Berlin, Heidelberg, 2001, pp. 87–94.

A. E. Hoerl and R. W. Kennard. "Ridge regression: Biased estima- tion for nonorthogonal problems". In: *Technometrics* 12.1 (1970).

M. Hoffman, D. Blei, and F. Bach. "On-line learning for latent Dirichlet allocation". In: *Neural Information Processing Systems* (2010).

D. Hopkins and G. King. "A method of automated nonparametric content analysis for social science". In: *American Journal of Political Science* 54.1 (2010), pp. 229–247.

D. Hopkins and G. King. *ReadMe: Software for Automated Content Analysis*. 2013. http://gking.harvard.edu/readme. R package version 0.99836.

A. Hotho, A. Nurnberger, and G. Paaß. *A Brief Survey of Text Mining*. LDV Forum, 2005.

S. M. Iacus, G. Porro, S. Salini, and E. Siletti. "Social networks, happiness and health: From sen- timent analysis to a multidimensional indicator of subjective well-being". In: *ArXiv e-prints* (Dec. 2015). arXiv: `1512.01569[stat.AP]`.

S. M. Iacus, G. King, and G. Porro. "Multivariate matching methods that are monotonic imbalance bounding". In: *Journal of the American Statistical Association* 106 (2011), pp. 345–361.

S. M. Iacus. *Simulation and Inference for Stochastic Differential Equations. With R Examples.* New York: Springer, 2008.

S. M. Iacus. "Big Data or big fail? The good, the bad and the ugly and the missing role of statistics". In: *Electronic Journal of Applied Statistical Analysis: Decision Support Systems and Services Evaluation* 5.1 (2014), pp. 4–11.

S. M. Iacus, G. Porro, S. Salini, and E. Siletti. "Controlling for selection bias in social media indicators through official statistics: A proposal". In: *Journal of Official Statistics* 36.2 (2020), pp. 315–338.

S. M. Iacus and N. Yoshida *Simulation and Inference for Stochastic Processes with YUIMA: A Comprehensive R Framework for SDEs and Other Stochastic Processes.* New York: Springer, 2018.

J. Iio. "Kawaii in Tweets: What Emotions Does the Word Describe in Social Media?" In: *Advances in Networked-based Information Systems.* Ed. by L. Barolli, H. Nishino, T. Enokido, and M. Takizawa. Cham: Springer International Publishing, 2020, pp. 715–721.

Istat. *Rapporto BES 2017. Il benessere equo e sostenibile in Italia.* Tech. rep. Istat, 2017. `https://www.istat.it/it/files//2017/12/Bes\_2017.pdf`.

K. Jaidka, S. Giorgi, H. A. Schwartz, M. L. Kern, L. H. Ungar, and J. C. Eichstaedt. "Estimating geographic subjective well-being from Twitter: A comparison of dictionary and data-driven language methods". In: *Proceedings of the National Academy of Sciences* 117.19 (2020), pp. 10165–10171.

Matthew L. Jockers. *Syuzhet: Extract Sentiment and Plot Arcs from Text.* 2015. `https://github.com/mjockers/syuzhet`.

N.M. Jones, S.P. Wojcik, J. Sweeting, and R.C. Silver. "Tweeting negative emotion: An investigation of Twitter data in the aftermath of violence on college campuses". In: *Psychological Methods* 21.4 (2016), pp. 526–541.

D. Kahneman, E. Diener, N. Schwarz (Eds). *Well-Being: The Foundations of Hedonic Psychology.* New York: Russell Sage Foundation, 1999.

D. Kahneman and A. Deaton. "High income improves evaluation of life but not emotional well-being". In: *Proceedings of the National Academy of Sciences* 107.38 (2010), pp. 16489–16493.

D. Kahneman and A. B. Krueger. "Developments in the measurement of subjective well-being". In: *Journal of Economic Perspectives* 20.1 (2006), pp. 3–24.

D. Kahneman, A. B. Krueger, D. Schkade, N. Schwarz, and A. Stone. "Toward national well-being ac- counts". In: *American Economic Review* 94.2 (2004), pp. 429–434.

K. Kamijo, T. Nasukawa, and H. Kitamura. "Personality Estimation from Japanese Text". In: *PEOPLES@COLING*. 2016.

A. Karpathy. *CS231n: Convolutional Neural Networks for Visual Recog- nition*. 2018. http://cs231n.github.io/neural-networks-1/.

M. L. Kern, G. Park, J. C. Eichstaedt, H. A. Schwartz, M. Sap, L. K. Smith, and L. H. Ungar. "Gaining insights from social media language: Methodologies and challenges". In: *Psychological Methods* 21.4 (2016), pp. 507–525.

G. King, O. Rosen, and M.A. Tanner. *Ecological inference: New Methodological Strategies*. New York: Cambridge University Press, 2004.

G. King. "Ensuring the data rich future of the social sciences". In: *Science* 331 (2011), pp. 719–721.

G. King. "Preface: Big Data is Not About the Data!" In: *Computational social science: Discovery and Prediction*. Ed. by R. M. Alvarez. Cambridge: Cambridge University Press, 2016. Chap. 1, pp. 1–10.

T. Kohonen. *Self-Organizing Maps*. Springer Series in Information Theory. NewYork: Springer, 2001.

M. Kosinski, D. Stillwell, and T. Graepel. "Private traits and attributes are predictable from digital records of human behavior". In: *Proceedings of the National Academy of Sciences* 110.15 (2013), pp. 5802–5805.

A.D.I. Kramer, J.E. Guillory, and J.T. Hancock. "Experimental evidence of massive-scale emotional contagion through social networks". In: *Proceedings of the National Academy of Sciences USA* 111.24 (2014), pp. 8788–8790.

A. D. I. Kramer. "An Unobtrusive Behavioral Model of Gross National Happiness". In: *Proceedings of the SIGCHI Conference on Human Factors in Computing Systems*. CHI '10. New York, USA: ACM, 2010, pp. 287– 290.

M. Kumano. "On the concept of well-being in Japan: Feeling shiawase as hedonic well-being and feeling ikigai as eudaimonic well-being". In: *Applied Research in Quality of Life* 13.2 (2018), pp. 419–433.

S. Kuznets. "National Income, 1929–1932". In: *National Income, 1929– 1932*. NBER Chapters. National Bureau of Economic Research, Inc., 1934, pp. 1–12.

H. A. Landsberger. *Hawthorne Revisited: Management and the Worker, Its Critics, and Developments in Human Relations in Industry*. Ithaca, NY: Cornell University, 1958.

S. Latouche. *Le pari de la décroissance*. Paris: Fayard, 2006.

M. Laver, K. Benoit, and J. Garry. "Extracting policy positions from political texts using words as data". In: *American Political Science Review* 97.2 (2003), pp. 311–331.

D. Lazer, A. Pentland, L. Adamic, S. Aral, A. L. Barabasi, D. Brewer, N. Christakis, N. Contractor, J. Fowler, M. Gutmann, T. Jebara, G. King, M. Macy, D. Roy, and M. Van Alstyne. "Computational social science". In: *Science* 323.5915 (2009), pp. 721–723.

J.A. Lee, C. Efstratiou, and L. Bai. "OSN Mood Tracking: Exploring the Use of Online Social Network Activity as an Indicator of Mood Changes". In: *Proceedings of the 2016 ACM International Joint Conference on Pervasive and Ubiquitous Computing: Adjunct*. ACM, 2016, pp.1171–1179.

K.A. Levin and C. Currie. "Reliability and validity of an adapted version of the Cantril Ladder for use with adolescent samples". In: *Social Indicators Research* 119.2 (2014), pp. 1047–1063.

A. Liaw and M. Wiener. "Classification and regression by random- Forest". In: *R News* 2.3 (2002), pp. 18–22. http://CRAN.R-project.org/doc/Rnews/.

K. H. Lim, K. E. Lee, D. Kendal, L. Rashidi, E. Naghizade, S. Winter, and M. Vasardani. "The Grass is Greener on the Other Side: Understanding the Effects of Green Spaces on Twitter User Sentiments". In: *Companion of the The Web Conference, 2018*. International World Wide Web Conferences Steering Committee. 2018, pp. 275–282.

B. Liu. *Web Data Mining; Exploring Hyperlinks, Contents, and Usage Data*. New York: Springer, 2006.

P. Liu, W. Tov, M. Kosinski, D. J. Stillwell, and L. Qiu. "Do Facebook status updates reflect subjective well-being?" In: *Cyberpsychology, Behavior, and Social Networking* 18.7 (2015), pp. 373– 379.

M. Luhmann. "Using Big Data to study subjective well-being". In: *Current Opinion in Behavioral Sciences* 18 (2017), pp. 28–33.

S. Lyubomirsky, L. King, and E. Diener. "The benefits of frequent positive affect: Does happiness lead to success?" In: *Psychological Bulletin* 131.6 (2005), pp. 803–855.

A. L. Maas, R. E. Daly, P. T. Pham, D. Huang, A. Y. Ng, and C. Potts. "Learning Word Vectors for Sentiment Analysis". In: *Proceedings of the 49th Annual Meeting of the Association for Com- putational Linguistics: Human Language Technologies*. Portland, Ore- gon, USA: Association for Computational Linguistics, 2011, pp. 142–150.

D. Marazziti, A. Pozza, M. Di Giuseppe, and C. Conversano. "The psychoso- cial impact of COVID-19 pandemic in Italy: A lesson for mental health prevention in the first severely hit European country". In: *Psychologi- cal Trauma: Theory, Research, Practice, and Pol- icy* 12.5 (2020), pp. 531–533.

S. Marchetti, C. Giusti, and M. Pratesi. "The use of Twitter data to improve small area estimates of households' share of food consumption expendi- ture in Italy". In: *AStA Wirtschafts - und Sozialstatistisches Archiv* 10.2 (2016), pp. 79–93.

S. Marchetti, C. Giusti, M. Pratesi, N. Salvati, F. Giannotti, D. Pedreschi, S. Rinzivillo, L. Pappalardo, and L. Gabrielli. "Small area model-based estimators using big data sources". In: *Journal of Official Statistics* 31.2 (2015), pp. 263–281.

Y. Marhuenda, I. Molina, and D. Morales. "Small area es- timation with spatio-temporal Fay-Herriot models". In: *Computational Statistics & Data Analysis* 58 (2013), pp. 308–325.

P. Massicotte and D. Eddelbuettel. "gtrendsR: Perform and display Google Trends queries". 2020. `https://CRAN.R-project.org/package=gtrendsR`.

D. Matsumoto. "American-Japanese cultural differences in judgements of ex- pression intensity and subjective experience". In: *Cognition and Emotion* 13.2 (1999), pp. 201–218.

G. Maugeri, P. Castrogiovanni, G. Battaglia, R. Pippi, V. D' Agata, A. Palma, M. Di Rosa, and G. Musumeci. "The impact of physical activity on psychological health during Covid-19 pandemic in Italy". In: *Heliyon* 6.6 (2020), pp. 1–8.

W. J. McKibbin and R. Fernando. "The Global Macroeconomic Impacts of Covid-19: Seven Scenarios". 2020. `https://www.brookings.edu/research/the-global-macroeconomic-impactsof-covid-19-seven-scenarios/`.

R.C. Merton. "Theory of rational option pricing". In: *The Bell Journal of Economics and Management Science* 4.1 (1973), pp. 141–183.

G. Mestre-Bach, G. R. Blycker, and M. N. Potenza. "Pornog- raphy use in the setting of the COVID-19 pandemic". In: *Journal of Be- havioral Addictions* 9.2 (2020), pp. 181–183.

N. Metropolis, A. W. Rosenbluth, M. N. Rosenbluth, A. H. Teller, and E. Teller. "Equation of state calculations by fast computing ma- chines". In: *The Journal of Chemical Physics* 21 (June 1953), pp. 1087– 1092.

D. Meyer, E. Dimitriadou, K. Hornik, A. Weingessel, and F. Leisch. *e1071: Misc Functions of the Department of Statistics (e1071), TU Wien.* 2014. http://CRAN.R-project.org/package=e1071. R package version 1.6-3.

A. C. Michalos, B. Smale, R. Labonté, N. Muharjarine, K. Scott, K. Moore, L. Swystun, B. Holden, H. Bernardin, B. Dunning, P. Graham, M. Guhn, A. M. Gadermann, B. D. Zumbo, A. Morgan, A. S. Brooker, and I. Hyman. *The Canadian Index of Wellbeing. Technical Report 1.0. Waterloo.* Tech. rep. UNESCO, 2011. http://www.unesco.org/new/fileadmin/MULTIMEDIA/ HQ/CLT/pdf/canadianindexofwellbeingtechnicalpaper.pdf.

T. Mikolov, K. Chen, G. Corrado, and J. Dean. "Efficient Estimation of Word Representations in Vector Space". In: *Proceedings of the ICLR Workshop 2013.* Computational and Biological Learning Society. 2013.

T. Mikolov, K. Chen, G. Corrado, and J. Dean. "Efficient Estimation of Word Representations in Vector Space". In: *Proceedings of Workshop at ICLR.* 2013.

T. Mikolov, I. Sutskever, K. Chen, G. Corrado, and J. Dean. "Distributed Representations of Words and Phrases and Their Compositionality". In: *Proceedings of NIPS.* 2013.

T. Mikolov, W.-t. Yih, and G. Zweig. "Linguistic Regularities in Continuous Space Word Representations". In: *Proceedings of NAACL HLT.* 2013.

T. Minka and J. Lafferty. "Expectation-Propagation for the Generative As- pect Model". In: *Proceedings of the 18th Conference on Uncertainty in Artificial Intelligence.* Elsevier, New York, 2002.

H. Minqing and L. Bing. "Mining and Summarizing Customer Reviews". In: *Proceedings of the ACM SIGKDD International Conference on Knowledge Discovery & Data Mining (KDD-2004).* 2004.

L. Mitchell, M. R. Frank, K. D. Harris, P. S. Dodds, and C. M. Danforth. "The geography of happiness: Connecting Twitter sen- timent and expression, demographics, and objective characteristics of place". In: *PLoS ONE* 8.5 (2013).

A. Miura, M. Komori, N. Matsumara, and K. Maeda. "Expression of negative emotional responses to the 2011 Great East Japan Earthquake: Analysis of big data from social media". In: *Japanese Journal of Psychology* 86.2 (2015), pp. 102–111.

K. Miyake. "How young Japanese express their emotions visually in mo- bile phone messages: A sociolinguistic analysis". In: *Japanese Studies* 27.1 (2007), pp. 53–72.

S. M. Mohammad and P. D. Turney. "Crowdsourcing a word-emotion asso- ciation lexicon". In: *Computational Intelligence* 29.3 (2013), pp. 436–465.

I. Molina and Y. Marhuenda. "sae: An R package for small area estimation". In: *The R Journal* 7.1 (2015), pp. 81–98. https://journal.r-project. org/archive/2015/RJ-2015-007/index.html.

F. Morstatter and H. Liu. "Discovering, assessing, and mitigating data bias in social media". In: *Online Social Networks and Media* 1 (2017), pp. 1–13.

I. Mukherjee and D. M. Blei. "Relative Performance Guarantees for Approx- imate Inference in Latent Dirichlet Allocation". In: *Advances in Neural Information Processing Systems 21*. Ed. by D. Koller, D. Schuurmans, Y. Bengio, and L. Bottou. 2009, pp. 1129–1136.

J. Murphy, M. W. Link, J. H. Childs, C. L. Tesfaye, E. Dean, M. Stern, J. Pasek, J. Cohen, M. Callegaro, and P. Harwood. "Social media in public opinion research. Executive sum- mary of the AAPOR task force on emerging technologies in public opinion research". In: *Public Opinion Quarterly* 78.4 (2014), pp. 788–794.

Office for National Statistics. *International Passenger Survey*. Available: http://doi.org/10.5255/UKDA-SN-8468-1. Sept. 2019.

New Economics Foundation. *The Happy Planet Index 2016. A Global Index of Sustainable Well-Being*. Tech. rep. New Eco- nomics Foundation, 2016. https://static1.squarespace.com/ static/5735c421e321402778ee0ce9/t/57e0052d440243730fdf03f3/ 1474299185121/Briefing+paper+-+HPI+2016.pdf.

New Zealand Ministry of Social Development. *The Social Report 2016*. Tech. rep. New Zealand Ministry of Social Development, 2016. http://socialreport.msd.govt.nz/documents/2016/ msd-the-social-report-2016.pdf.

M. Nicola, Z. Alsafi, C. Sohrabi, A. Kerwan, A. Al-Jabir, C. Iosifidis, M. Agha, and R. Agha. "The socio-economic implications of the coronavirus pandemic (COVID-19): A review." In: *International journal of surgery* 78 (2020), pp. 185–193.

W. Nordhaus and J. Tobin. "Is Growth Obsolete?" In: *The Measure- ment of Economic and Social Performance*. National Bureau of Economic Research, Inc., 1973, pp. 509–564.

P. K. Novak, J. Smailović, B. Sluban, and I. Mozetič. "Sentiment of Emojis". In: *PLoS ONE* 10.12 (2015), pp. 1–22.

S.V. Nuti, B. Wayda, I. Ranasinghe, S. Wang, R. P. Dreyer, S. I. Chen, and K. Murugiah. "The use of Google Trends in health care research: A system- atic review". In: *PLoS ONE* 10.9 (2014), pp. 1–30.

OECD. *How's Life? 2013. Measuring Well-being*. OECD Publishing, 2013. `http://dx.doi.org/10.1787/9789264201392-en`.

OECD. *How's Life? 2017. Measuring Well-being*. OECD Publishing, 2017. `https://doi.org/10.1787/how\_life-2017-en`.

K. Omori. "Cultural Differences in Self-Presentation on Social Network- ing Sites: A Cross-cultural Comparison Between American and Japanese College Students". PhD thesis, University of Wisconsin-Milwaukee, 2014.

M. Orgilés, A. Morales, E. Delvecchio, C. Mazzeschi, and J. P. Espada. "Immediate psychological effects of the Covid-19 quarantine in youth from Italy and Spain". 2020. `https://doi.org/10.31234/osf.io/5bpfz`.

B. Pang and L. Lee. "A Sentimental Education: Sentiment Analysis Using Subjectivity". In: *Proceedings of the ACL-04 Conference on Empirical Methods in Natural Language Processing*. 2004, pp. 271–278.

B. Pang, L. Lee, and S. Vaithyanathan. "Thumbs Up?: Sentiment Classification Using Machine Learning Techniques". In: *Proceedings of the ACL-02 Conference on Empirical Methods in Natural Language Processing*. 2002, pp. 79–86.

G. Panger. "Reassessing the Facebook experiment: Critical thinking about the validity of Big Data research". In: *Information, Communication & Society* 19.8 (2016), pp. 1108–1126.

J. Park, Y. M. Baek, and M. Cha. "Cross-cultural com- parison of nonverbal cues in emoticons on Twitter: Evidence from Big Data analysis". In: *Journal of Communication* 64.2 (2014), pp. 333–354.

J. Pearl. "Causal diagrams for empirical research". In: *Biometrika* 82.4 (1995), pp. 669–688.

J. Pearl and S. Russell. "Bayesian Network". In: *The Handbook of Brain Theory and Neural Networks*. Ed. by M. A. Arbib. Cambridge, MA: MIT Press, 2003, pp. 157–160.

J. W. Pennebaker, M. E. Francis, and R. J. Booth. *The Development and Psychometric Prop- erties of LIWC2015*. University of Texas at Austin, 2015.

J. W. Pennebaker, M. E. Francis, and R. J. Booth. *Linguistic Inquiry and Word Count*. Mahwah, NJ: Lawerence Erlbaum Associates, 2001.

A. Pentland. *Social Physics: How Good Ideas Spread-the Lessons from a New Science*. Penguin Publishing Group, 2014.

R. Plutchik. *Emotion, a Psychoevolutionary Synthesis*. New York: Harper & Row, 1980.

R. Plutchik. "The nature of emotions: Human emotions have deep evo- lutionary roots, a fact that may explain their complexity and provide tools for clinical practice". In: *American Scientist* 89.4 (2001), pp. 344–350.

A. T. Porter, S. H. Holan, C. K. Wikle, and N. Cressie. "Spatial Fay-Herriot models for small area estimation with functional covariates". In: *Spatial Statistics* 10 (2014), pp. 27–42.

M. Ptaszynski, R. Rzepka, K. Araki, and Y. Momouchi. "Automatically annotating a five-billion-word corpus of Japanese blogs for sentiment and affect analysis". In: *Computer Speech & Language* 28.1 (2014), pp. 38–55.

K. Qian and T. Yahara. "Mentality and behavior in COVID-19 emergency status in Japan: Influence of personality, morality and ideology". In: *PLoS ONE* 15.7 (2020), pp. 1–16.

D. Quercia. "Hyperlocal Happiness from Tweets". In: *Twitter: A Digital Socioscope*. Ed. by Y. Mejova, I. Weber, and M. W. Macy. Cambridge University Press, 2015, pp. 96–110.

D. Quercia, J. Ellis, L. Capra, and J. Crowcroft. "Tracking Gross Community Happiness from Tweets". In: *Proceedings of the ACM 2012 Conference on Computer Supported Cooperative Work*. CSCW '12. ACM, 2012, pp. 965–968.

K. Quinn. "How to analyze political attention with minimal assumptions and costs". In: *American Journal of Political Science* 54.1 (2010), pp. 209–228.

R Core Team. "R: A Language and Environment for Statistical Computing". Vienna, Austria, 2020. https://www.R-project.org/.

J. N. K. Rao and M. Yu. "Small-area estimation by combining time- series and cross-sectional data". In: *The Canadian Journal of Statistics* 22.4 (1994), pp. 511–528.

C. P. Robert and G. Casella. *Monte Carlo Statistical Methods (Springer Texts in Statistics)*. Berlin, Heidelberg: Springer-Verlag, 2005.

M. E. Roberts, B. M. Stewart, D. Tingley, C. Lucas, J. Leder-Luis, S. K. Gadarian, B. Albertson, and D. G. Rand. "Structural topic models for open-ended survey re- sponses". In: *American Journal of Political Science* 58.4 (2014), pp. 1064– 1082.

I. Robeyns. *An Unworkable Idea or a Promising Alternative? Sen's Ca- pability Approach Re-examined.* Center for Economic Studies. Discussions Paper Series n. 30.2000.

I. Robeyns. "The capability approach: A theoretical survey". In: *Journal of Human Development* 6.1 (2005), pp. 93–117.

I. Robeyns. "The capability approach in practice". In: *Journal of Political Philosophy* 14.3 (2006), pp. 351–376.

C. R. Rogers. *On Becoming a Person: A Therapist's View of Psychotherapy.* London: Constable, 1961.

P. R. Rosembaum and D. B. Rubin. "The central role of the propen- sity score in observational studies for causal effects". In: *Biometrika* 70.1 (1983), pp. 41–55.

Y. Rosseel. "lavaan: An R package for structural equation modeling". In: *Journal of Statistical Software* 48.2 (2012), pp. 1–36. http://www.jstatsoft.org/v48/i02/

R. Rossi, V. Socci, D. Talevi, S. Mensi, C. Niolu, F. Pacitti, A. Di Marco, A. Rossi, A. Siracusano, and G. Di Lorenzo. "Covid-19 pandemic and lock- down measures impact on mental health among the general population in Italy". In: *Frontiers in Psychiatry* 11.790 (2020), pp. 1–6.

S. Rossouw and T. Greyling. "Big data and happiness". In: *Hand- book of Labor, Human Resources and Population Economics.* Ed. by K. F. Zim- mermann. Springer, 2020, pp. 1–35.

D. Ruths and J. Pfeffer. "Social media for large studies of behavior". In: *Science* 346.6213 (2014), pp. 1063–1064.

J. A. Ryan and J. M. Ulrich. "quantmod: Quantitative Financial Modelling Framework". 2020. https://CRAN.R-project.org/package=quantmod.

R. M. Ryan and E. L. Deci. "On happiness and human potentials: A review of research on hedonic and eudaimonic well-being". In: *Annual Review of Psychology* 52.1 (2001), pp. 141–166.

R. Rzepka, U. Jagla, P. Dybala, and K. Araki. "Influence of Emoticons and Adverbs on Affective Perception of Japanese Texts". In: *Proceedings of the 30th Annual Conference of the Japanese Society for Artificial Intelligence* JSAI2016.1 (2016).

C. Salvatore, S. Biffignandi, and A. Bianchi. "Social media and Twitter data quality for new social indicators". In: *Social Indicators Research* (2020), pp. 1–30.

G. Sani, D. Janiri, M. Di Nicola, L Janiri, S. Ferretti, and D. Chieffo. "Mental health during and after the Covid-19 emergency in Italy". In: *Psychiatry and Clinical Neurosciences* 74.6 (2020), p. 372.

J. Schmidhuber. "Deep Learning in Neural Networks: An Overview". In: *CoRR* abs/1404.7828 (2014). arXiv: 1404 . 7828.

J. Schork, C. A. F. Riillo, and J. Neumayr. *Survey Mode Ef- fects on Objective and Subjective Questions: Evidence from the Labour Force Survey.* Tech. rep. n. 109 Économie et Statistiques. Working Papers du STATEC, 2019.

H. A. Schwartz, J. C. Eichstaedt, M. L. Kern, L. Dziurzynski, M. Agrawal, G. J. Park, S. K. Lakshmikanth, S. Jha, M. E. P. Seligman, L. Ungar, and R. E. Lucas. "Characterizing Geographic Variation in Well-being Using Tweets". In: *Proceedings of the Seventh International AAAI Conference on Weblogs and Social Media (ICWSM).* 2013.

H. A. Schwartz, M. Sap, M. L. Kern, J. C. Eichstaedt, A. Kapelner, M. Agrawal, and L. H. Ungar. "Predicting individual well-being through the language of social media". In: *Pacific Symposium On Biocomputing* 21 (2016), pp. 516–527.

N. Schwarz. "Self-reports: How the questions shape the answers". In: *American Psychologist* 54.2 (1999), pp. 93–105.

N. Schwarz and F. Strack. "Reports of subjective well-being: Judg- mental processes and their methodological implications". In: *Well-being: The Foundations of Hedonic Psychology* 7 (1999), pp. 61–84.

C. N. Scollon. "Non-traditional Measures of Subjective Well-being and Their Validity: A review". In: *Handbook of Well-being.* Ed. by E. Diener, S. Oishi, and L. Tay. Salt Lake City, UT: DEF Publishers, 2018.

J.P. Seder and S. Oishi. "Intensity of smiling in Facebook photos predicts future life satisfaction". In: *Social Psychological and Personality Science* 3.4 (2012), pp. 407–413.

A. Sen and G. Hawthorn. *The Standard of Living.* Tanner Lectures in Human Values. Cambridge University Press, 1988.

A. Sen. "Equality of What?" In: *The Tanner Lectures on Human Values.* Ed. by Sterling M. MacMurrin. Vol. 1. Cambridge: Cambridge University Press, 1980, pp. 195–220.

A. Sen. *Inequality Reexamined.* Oxford: Clarendon Press, 1992.

A. Sen. "Development as Capability Expansion". In: *Readings in Hu- man Development*. Ed. by S. Fukuda-Parr and A. K. Shiva Kumar. New Delhi and New York: Oxford University Press, 2003. Chap. 1, pp. 1–10.

A. Sen. "Capability and Well-being". In: *The Philosophy of Economics: An Anthology*. Ed. by D. Hausman. New York: Cambridge University Press, 2008, pp. 270–294.

J. Shigemura and M. Kurosawa. "Mental Health Impact of the COVID-19 Pandemic in Japan". In: *Psychological Trauma: Theory, Research, Practice, and Policy* 12.5 (2020), pp. 478–479.

J. Shigemura, R. J. Ursano, J. C. Morganstein, M. Kurosawa, and D. M. Benedek. "Public responses to the novel 2019 coron- avirus (2019-nCoV) in Japan: Mental health consequences and target pop- ulations". In: *Psychiatry and Clinical Neurosciences* 74.4 (2020), pp. 281– 282.

A. Shoeb and G. de Melo. "Are emojis emotional? A study to understand the association between emojis and emotions". ArXiv: 2005.00693.

B. B. Singh, G. K. Shukla, and D. Kundu. "Spatio-temporal models in small area estimation". In: *Survey Methodology* 31.2 (2005), pp. 183–195.

J.B. Slapin and S.-O. Proksch. "A scaling model for estimating time-series party positions from texts". In: *American Journal of Political Science* 52.3 (2008), pp. 705–772.

E. Soukiazis and S. Ramos. "The structure of subjective well-being and its determinants: A micro-data study for Portugal". English. In: *Social Indicators Research* 126.3 (2016), pp. 1375–1399.

S. Spyratos, M. Vespe, F. Natale, S. M. Iacus, and C. Santamaria. "Explaining the travelling behaviour of migrants using Facebook audience estimates". In: *PLoS ONE* 15.9 (2020), pp. 1–16.

S. Spyratos, M. Vespe, F. Natale, I. Weber, E. Zagheni, and M. Rango. "Quantifying international human mobility patterns using Facebook network data". In: *PLOS ONE* 14.10 (2019), pp. 1–22.

J. K. Stanley, D. A. Hensher, J. R. Stanley, and D. Vella-Brodrick. "Mobility, social exclusion and well-being: Exploring the links". In: *Transportation Research Part A: Policy and Practice* 45.8 (2011), pp. 789–801.

S. Stephens-Davidowitz. *Everybody Lies: Big Data, New Data, and What the Internet Can Tell Us About Who We Really Are*. New York, NY: William Morrow & Co, 2018.

A. Steptoe, A. Deaton, and A. A. Stone. "Subjective wellbeing, health, and ageing". In: *The Lancet* 385.9968 (2015), pp. 640–648.

D. H. W. Steyn, T. Greyling, S. Rossouw, and J. M. Mwamba. *Sentiment, emotions and stock market predictability in developed and emerging markets*. GLO Discussion Paper 502. Global Labor Organization, 2020.

J. Stiglitz, A. Sen, and J.-P. Fitoussi. *Report by the Commis- sion on the Measurement of Economic Performance and Social Progress*. Tech. rep. INSEE, 2009.

A. Strehl and J. Ghosh. "Cluster ensembles - a knowledge reuse framework for combining multiple partitions". In: *Journal of Machine Learning Research* 3 (2002), pp. 583–617.

S. Sun, J. Chen, M. Johannesson, P. Kind, and K. Burström. "Subjective well-being and its association with subjective health status, age, sex, region, and socio-economic characteristics in a Chinese population study". In: *Journal of Happiness Studies* 17(2016), pp. 833–873.

Y. Teh, M. Jordan, M. Beal, and D. M. Blei. "Hierarchical Dirichlet processes". In: *Journal of the American Statistical Association* 101.476 (2006), pp. 1566–1581.

The Human Development Report Office. *Human Development Report. Human Development for Everyone*. Tech. rep. UNDP, 2016. http://hdr.undp.org/en/2016-report.

R. Tibshirani. "Regression shrinkage and selection via the Lasso". In: *Journal of the Royal Statistical Society. Series B* 58 (1996), pp. 267–288.

L. Torricelli, M. Poletti, and A. Raballo. "Managing Covid-19 related psychological distress in health workers: Field experience in northern Italy". In: *Psychiatry and Clinical Neurosciences* 75.1, (2021), pp. 23–24.

J. Turian, L. Ratinov, and Y. Bengio. "Word Representations: A Simple and General Method for Semi-supervised Learning". In: *Proceedings of the 48th Annual Meeting of the Association for Computational Linguistics, ACL'10*. Stroudsburg, PA: Association for Computational Linguistics, 2010, pp. 384–394.

M. Ueda, A. Stickley, H. Sueki, and T. Matsubayashi. "Mental health status of the general population in Japan during the COVID-19 pandemic". In: *Psychiatry and Clinical Neurosciences* 74.9 (2020), pp. 505–506.

G.E. Uhlenbeck and L.S. Ornstein. "On the theory of Brownian motion". In: *Physical Review* 36 (1930), pp. 823–841.

K. Ujjwal. *A Quick Introduction to Neural Networks*. 2016. https://ujjwalkarn.me/2016/08/09/quick-intro-neural-networks/.

United Nations Department of Economic and Social Affairs. *Population by 5-year Age Groups, Annually from 1950 to 2100: Medium Projection Variant*. Available: `https://esa.un.org/unpd/wpp/Download/Standard/CSV/`. Mar. 2017.

United Nations Development Programme. *Human Development Report 2019. Beyond Income, beyond Averages, beyond Today: Inequalities in Human Development in the 21st Century*. Tech. rep. United Nations Development Programme, 2019. `http://hdr.undp.org/en/content/human-development-report-2019`

W. van der Wielen and S. Barrios. "Economic sentiment during the COVID pandemic: Evidence from search behaviour in the EU". In: *Journal of Economics and Business* 1 (2020), pp. 1–10.

O. Vasicek. "An equilibrium characterization of the term structure". In: *Journal of Financial Economics* 5 (1977), pp. 177–188.

R. Veenhoven. "Capability and happiness: Conceptual difference and reality links". In: *Journal of Behavioral and Experimental Economics (formerly The Journal of Socio-Economics)* 39.3 (2010), pp. 344–350.

D. A. Vella-Brodrick and J. Stanley. "The significance of transport mobility in predicting well-being". In: *Transport Policy* 29 (2013), pp. 236–242.

B.-K H. Vo and N. Collier. "Twitter emotion analysis in earthquake situations". In: *International Journal of Computational Linguistics and Applications* 4.1 (2013), pp. 159–173.

V. Voukelatou, L. Gabrielli, I. Miliou, S. Cresci, R. Sharma, M. Tesconi, and L. Pappalardo. "Measuring objective and subjective well-being: Dimensions and data sources". In: *International Journal of Data Science and Analytics* (2020), pp. 1–31.

L. C. Walz, M. H. Nauta, and M. A. H. Rot. "Experience sampling and ecological momentary assessment for studying the daily lives of patients with anxiety disorders: A systematic review". In: *Journal of Anxiety Disorders* 28.8 (2014), pp. 925–937.

C. Wang, D. M. Blei, and D. Heckerman. "Continuous time dynamic topic mod- els". In: *Proceedings of the Twenty-Fourth Conference on Uncertainty in Artificial Intelligence*. July 2008. 2008, pp. 579–586.

N. Wang, M. Kosinski, D. J. Stillwell, and J. Rust. "Can well-being be measured using Facebook sta- tus updates? Validation of Facebook's Gross National Happiness Index". In: *Social Indicators Research* 115.1 (2014), pp. 483–491.

Z. Wang, Z. Yu, R. Fan, and B. Guo. "Correcting biases in online social media data based on target distributions in the physical world". In: *IEEE Access* 8 (2020), pp. 15256–15264.

K. Watanabe. "Measuring Bias in International News: A Large-scale Analysis of News Agency Coverage of the Ukraine Crisis". PhD thesis. The London School of Economics and Political Science (LSE), 2017.

Canadian Index of Wellbeing. *How Are Canadians Really Doing? The 2016 CIW National Report.* Tech. rep. University of Waterloo, 2016. https://uwaterloo.ca/canadian-index-wellbeing/sites/ca.canadian-index-wellbeing/files/uploads/files/c011676-nationalreport-ciw_final-s_0.pdf.

Wikipedia. *Support Vector Machine.* 2018.

R. Winkelmann. "Unhappiness and unemployment". In: *IZA World of Labor* 94 (2014).

I.H. Witten. *Text Mining, Practical Handbook of Internet Computing.* Boca Raton, FL: CRC Press, 2004.

T. Yamamoto, C. Uchiumi, N. Suzuki, J. Yoshimoto, and E. Murillo-Rodriguez. "The psychological impact of 'mild lockdown'in Japan during the COVID-19 pandemic: A nationwide survey under a declared state of emergency". In: *International Journal of Environmental Research and Public Health* 17.24 (2020), 9382.

C. Yang and P. Srinivasan. "Life satisfaction and the pursuit of happiness on Twitter". In: *PLoS ONE* 11.3 (2016), pp. 1–30.

L. M. R. Ybarra and S. L. Lohr. "Small area estimation when auxiliary information is measured with error". In: *Biometrika* 95.4 (2008), pp. 919–931.

E. Zagheni, I. Weber, and K. Gummadi. "Leveraging Face- book's advertising platform to monitor stocks of migrants". In: *Popula- tion and Development Review* 43.4 (2017), pp. 721–734.

M. Zappavigna. *Discourse of Twitter and Social Media: How We Use Language to Create Affiliation on the Web.* New York: Continuum, 2012.

Y. Zhao, F. Yu, B. Jing, X. Hu, A. Luo, and K. Peng. *An Analysis of Well-Being Determinants at the City Level in China Using Big Data.* In: *Social Indicators Research* 143 (2019), pp. 973-994.

H. Zou and T. Hastie. "Regularization and variable selection via the elastic net". In: *Journal of the Royal Statistical Society. Series B* 67.2 (2005), pp. 301–320.

Index